THE
AMERICAN
FARM

THE AMERI

HOW FARMING SHAPED THE LANDSCAPE OF AMERICA

CAN FARM

John Fraser Hart

Adapted by Gail Kinn

B&N BOOKS
NEW YORK

1998 Barnes & Noble Books

ISBN 0-7607-0614-X

Book design by Patti Ratchford

Printed and bound in China

98 99 00 01 02 M 9 8 7 6 5 4 3 2 1

lfa

CONTENTS

◆◘◆

PREFACE

Taking the kids to visit Grandfather's farm used to be one of our great national traditions, but today's grandparents live in a town or city, just like the rest of us. As a remedy—of sorts—to our loss and our lack of exposure today to the magnificent landscape of American farming, this book not only reveals what farmers do, how and why they do it, but also shows how their efforts affect the rural landscape of America. ◨ Our route traces the mainstream of American agriculture westward from the seedbed in southeastern Pennsylvania to the western edge of the Corn Belt in southwestern Minnesota, and then it swings back along the Milky Way of dairy country from Wisconsin to New England, with a glance at the fruit-farming area of southwestern Michigan along the way. From New England the route turns along the Atlantic Coast to the plains land South, a region with quite a different agricultural tradition, and it visits some of the "islands" of agricultural specialization within the vast sea of dark pine forest that covers most of the region. ◨ Farming was the great engine that drove

American life from 1620 until 1860, perhaps until 1940, when one of every four Americans still lived on a farm. It was a very diverse engine that reflected the environmental variability of a continent. That diversity became even greater when different groups of people with different values and backgrounds identified their individual opportunities in similar environments and responded to them in different ways. The basic American farming system was developed in southeastern Pennsylvania, whence it crossed the Appalachian Uplands barrier to flourish in the Corn Belt. In the dairy areas along the northern fringe of the Corn Belt it was modified in response to environmental constraints. A few well-situated farmers produced milk, fruit, and truck crops for sale to city people, but for most American farmers in the nineteenth century, farming was a way of life rather than a business. They were content to produce enough to feed and clothe their families, with some slight surplus they could sell to buy the goods they could not make or grow on the farm. ◨ The westward march of this farming system

finally ground to a halt in Kansas and Nebraska, where the rainfall was too scanty and too unreliable for any crop except wheat. West of Kansas even wheat was too risky, and farming gave way to ranching. Wheat farming and ranching were commercial activities from the very beginning. Farmers specialized in producing large quantities of a few commodities for distant markets. In the oases of the West a whole new commercial farming system developed that reached its peak in the Central Valley of California. Farmers in the West had to specialize in large-scale production of high-value crops that could stand the cost of transportation to distant markets. Farming in the West is so distinctive that it deserves separate treatment. ◈ The South is explored here, however, because the South has always had its own curious mixture of commercial and way-of-life farming. Many small farms in the South have done little more than support a meager existence, but farming has always been a business on the large farms that have specialized in producing cotton, sugar, rice, and other cash crops. ◈ After World War I, farming in the eastern United States began to change from a way of life to a business, and this change has accelerated since World War II. The successful modern family farm has had to become an efficient business; farmers who used to think in hundreds now must think in thousands. They have had to specialize and eliminate their less profitable activities. They have had to learn to manage money and machinery as skillfully as they manage crops and livestock; ulcers have replaced blisters as their main occupational complaint. ◈ The best way to understand farms is to talk to the people who are trying to make a living by farming. You will hear their voices here as they speak of their experiences, their determination, and their concerns. ◈ Farmers do not operate in a vacuum. They do things in certain ways because their fathers and grandfathers taught them the best way, but they must also cope with rapid economic and technological change, and they may have to adapt the old ways to new circumstances.

John Fraser Hart

Birthplace of the American Farm

AMERICAN FARMING, with the help of Europeans, all began here on the Lancaster Plain, which local people still call the Garden Spot. It remains one of the finest farming areas in North America.

The roots of American agriculture lie deep in the soils of Europe. In fact, many of the earliest seeds planted in America were imported from Europe. The Lancaster Plain of southeastern Pennsylvania, underlain by massive beds of limestone that form deep, rich, productive soils, provided one of the finest farming areas in North America. These fertile plains—the only extensive tract of truly good farming land on the eastern seaboard of the United States—provided the seedbed for early crops; later, the plants were transplanted to other parts of the nation. Summers were hotter and winters colder than in Europe, but both were tolerable. The growing season, between the last killing frost in mid-April and the first killing frost in mid-October, was six months long. Settlers had plenty of

time to plant and harvest. The land was well watered. Three to four inches of rain fell each month during the growing season, and droughts were rare. Even many of the plants and animals reminded the new settlers of home. Here they recognized oak, maple, ash, elm, and beech trees in the hardwood forests that covered the land, and they soon learned the value of hickory, walnut, chestnut, and tulip trees.

The trees gave them logs for buildings, rails for fences, and wood for their fireplaces. Trees were also useful indicators of the quality of the land, because different kinds of trees grow in different

kinds of soil. The settlers used the trees to identify the best farming areas. They learned to "read" the trees as easily as modern Americans read road signs. The soils along the coast were only middling, but west of Philadelphia was the Lancaster Plain, one of the most fertile regions. Local people still call it the Garden Spot. The rolling surface of the soil is easy to cultivate, and erosion does not pose any problem. The early settlers planted and harvested the familiar grain crops of wheat, oats, rye, and barley, and the new grain crop, corn, which, among other crops, the Native Americans taught them how to grow.

WHEAT WAS THE FIRST commercial crop grown in Pennsylvania. Within a generation their harvest would fill wagons with surplus grain to be shipped to Europe. Pictured above is an early farmyard in Germantown, Pa., circa 1900–1906.

(Right) The contemporary Lancaster Plain is a showplace, aesthetically as well as technologically. Many prosperous farmers were able to build lovely stone or brick structures with ten to fourteen rooms or more on their farms, which have been in the same families since 1750 or earlier.

In 1710 a large group of German Mennonites began farming in the Conestoga Valley near the present city of Lancaster. Wheat was the first commercial crop, as it was in many pioneer areas. Within a generation, farmers were down-hauling wagons loaded with grain to Philadelphia or Baltimore for shipment to Europe, and by 1757, French farmers were complaining that they could not grow wheat cheaply enough to compete with the quantity of imports from Pennsylvania. One third of the settlers' cultivated land was used to grow wheat, the rest to grow corn for cattle and hogs, oats for the horses, and rye for making whiskey. Every farm had a vegetable garden and an orchard of apple, peach, and cherry trees.

A farmer needed three or four horses to plow his land, and a team of four to pull a loaded Conestoga wagon to Philadelphia or Baltimore. Most farmers kept half a dozen motley cattle for milk and meat, fattened a few hogs, and had a flock of barnyard hens for eggs and Sunday dinner. They consumed about half of their produce on the farm and sold the rest either locally or for export.

After the American Revolution, when European farmers were caught up in the ferment of the Agricultural Revolution, many of their new ideas filtered across the Atlantic Ocean and changed nearly every aspect of farming. New crops were introduced and grown in new ways. Lancaster farmers began to

EXCEPT FOR AN OCCASIONAL strand of electrified wire, farmsteads are laced together by a maze of narrow paved roads that zigzag through the countryside. At the end of summer the rows of tall corn crowd right up to the edge of the pavement to create blind corners. If you wander from the main highways, it is woefully easy to get lost.

(Left) A worker relaxes atop a brand new hand-built wooden silo under construction, still being held up by the wire on the left.

grow clover as a soil-enrichment crop and used it for pasture or cut it for hay. Turnips, which are fodder roots, began to be used in the new agriculture. They were planted in rows, which controlled weeds, and were fed to the livestock. Instead of turnips, Pennsylvania farmers planted corn, the native American crop. Since World War II, two-thirds of the cropland in Lancaster County was planted with corn. Corn, like turnips, also provided excellent feed for fattening cattle and hogs; fattened cattle, in turn, produced prodigious quantities of manure that were returned to the soil. A quarter of the crop system is in hay, and most of the rest is in wheat or tobacco.

After each of America's major wars, farmers on the Lancaster Plain tried to build up their small farms. After the Revolutionary War they shifted from wheat farming to cattle fattening. After the Civil War ended, they began growing tobacco. After World War I, they switched from beef to dairy cattle as they could make more money selling milk. They added poultry and hog enterprises after World War II. By 1982, sales of poultry and eggs became a third of the county's total farm income, and sales of hogs added another 10 percent.

The contemporary Lancaster Plain is a showplace, aesthetically as well as technologically. Some farms have been

Crop Rotation

Crop Rotation The Lancaster Plain was instrumental in introducing new ways of farming to the rest of America. These consisted of the four-year crop rotation system; the mixed farming system, in which livestock and agriculture were farmed together; and distinctive barns to house the new farming techniques. All were transplanted to southwestern Ohio, where they were slightly modified and became the norm for the agricultural heartland of the continent.

THREE-YEAR ROTATION

The medieval three-year crop rotation of food, feed, and fallow land had remained virtually unchanged in Europe since the time of the Roman Empire; some European farmers were only reluctantly abandoning the ancient and primitive cropping system by the mid-eighteenth century. But the old system had led to overuse of land and massive infestations of weeds resulting in soil depletion. The only method of restoring the soil at the time, was to let it lie fallow for an entire year, leaving farmers and their families hungry and poor. The discovery of new farming methods and crops during the Agricultural Revolution changed this dramatically.

THE VIRTUES OF FOUR-YEAR ROTATION

◈ New crops enabled farmers to expand the old three-year rotation into a new and better four-year rotation.

◈ Farmers no longer needed to let a third of their land lie fallow each year both because they could control weeds by cultivating the root crop, and because the clover enriched the soil for the vital crop of food grain that followed it.

◈ The elimination of the fallow year increased the amount of land under crops by 50 percent, just when the Industrial Revolution caused a rapid escalation in the demand for food.

in the same family since 1750 or earlier, and some buildings date back to the Revolutionary War. Many farmhouses are massive stone or brick structures with ten to fourteen rooms or more. In front of the house is a neatly trimmed lawn and a luxuriant flower bed ablaze with color, and behind is a garden full of vegetables. But today, small farmers in Lancaster County are scratching for more land. The only way a young person can get started in farming is to inherit a farm or to marry into one. Many farmers have formed family partnerships or corporations to enable their children to become farmers.

Modern farming has essentially become big business. Industrial farmers have adopted a whole new technology, with better crops, animals, and buildings, and bigger debts and more sleepless nights to show for it. Conservationists have been concerned about the loss of agricultural land to industrial development and urban sprawl. Factories and stores appear at unexpected places along the country roads, and some farmers have sold narrow roadside strips for new nonfarm developments. Perhaps the best way to preserve land for agriculture is to sell it to an Amishman because you know he will never sell it.

THE STILLNESS OF A LANCASTER FARM in the early morning light obscures our awareness of the relentless hard work and nonstop activity taking place inside every building.

(Overleaf) Because of the fertile Pennsylvania soil, the landscape is a veritable tapestry of farms.

A Personal History

DON HERSHEY

MASTER FARMER IN PENNSYLVANIA COUNTRY

Don Hershey's farm is ten miles northwest of Lancaster, outside Amish country. A sturdy weatherbeaten sign beside the main road has a picture of a black-and-white cow on a green pasture, and beneath it "Hershvale Farms—Don, Gerry, Larry, Stephen & Patti." His family still lives in the house where Gerry, his wife, grew up.

"In 1975 I bought a second 100-acre farm. It was for sale and I had to buy it, even though I wasn't really ready, but I had to look ahead to the time when my sons would be joining me in the farm business. I had to expand the operation because Larry, born in 1959, and Steve, born in 1961, wanted to join me on the farm when they finished high school. I'm proud that they wanted to farm with me, but we had to have a bigger operation to support three families. I bought a second farm; it would have been a whole lot cheaper to send the boys to college. I'm basically a dairyman, but most of my expansion had to be in poultry and hogs, not cows."

Every member of the family helps out with everything, but each one has special responsibilities. Don runs the daily operation. He and Larry milk 80 cows in the barn at the home place, and Steve milks 30 more at the rented place where he lives.

Steve is the egg man. The poultry house has 31,000 laying

"I'M PROUD THAT [MY SONS] WANTED TO FARM WITH ME."

hens in wire-mesh cages stacked three high. "We do our best to keep the birds as happy as possible, because happy birds produce more eggs and make more money for us. It takes us about six hours a day to pack the eggs. We sell them to a local company. We expect a bird to produce nine eggs every ten days for twelve four-week laying periods. We induce molting for four weeks of rest and recovery, which makes the birds stop laying and start shedding their feathers."

Larry is the hog man. "We buy lean feeder pigs from a dealer when they are ten weeks old and weigh 40 pounds. We fatten them to a market weight of 220 pounds in 120 days."

In 1978, *Pennsylvania Farmer* magazine designated Don a Master Farmer, one of only six in a five-state area, but even a Master Farmer is at the mercy of prices, interest rates, the weather, and other forces over which he has no control. "In 1981, hogs, steers, and chicken prices all went bad; in 1982, interest went way up; and in 1983, I got hit by drought. The corn never even came to tassel; it just curled up."

Despite their vicissitudes, however, or perhaps because of them, the Hersheys have deep faith and the same reverence for the land that Lancaster County farmers have shared for two centuries. Nothing can compete with that. ◆

Barn Style for Mixed Farming

In the days before chemical fertilizer had been invented, you could tell how good a farmer was by the size of his manure pile. Farmers valued their animals for their manure almost as much as for their meat because they needed manure to fertilize their soil to grow crops to feed their animals. This interdependence of crops and livestock in a tightly integrated farming system was called mixed farming.

Mixed farming required new structures. The farmer not only had to protect his fields with fences, hedges, or stone walls, he also had to shelter the animals against harsh weather. He needed stout pens where he could store his crops, feed the animals, and collect their manure. Finally, he needed a barn where he could thresh and store his grain. Some Swiss farmers satisfied all their requirements under a single roof by building two-story structures with livestock on the lower level and a threshing floor on the upper level. These two-story "Swiss" barns were models for the new barns of southeastern Pennsylvania. They were a remarkable innovation and have become so widely accepted that most Americans take these efficient structures for granted.

TODAY LARGE CYLINDRICAL SILOS tower beside most barns, but they were not added until a century or so after the barn had been built, often when the farmer switched from beef to dairy cattle.

22

PUTTING UP A PERMANENT BARN was an exhilarating moment in the life of young family farmers, a sign that crops and livestock were thriving and all was going well.

(Bottom) Farmers pump water for washing and drinking during a break from their endless chores.

AROUND 1780, FARMERS on the Lancaster Plain began erecting the large, two-story "Swiss" barns so distinctive of the rural landscape of southeastern Pennsylvania. In front of this sparkling bright red example stands a gleaming black Amish buggy.

(Left) Today's farm is a veritable village of farm buildings with old and new silos for storing grains; multilevel wagonwide double-door barns; and a farmhouse, among other sheds and storage buildings. More than likely there will also be a host of dairy cows grazing on verdant pastureland.

Few Americans are aware of the fact that livestock have never been allowed in most European barns; instead, the animals are kept in separate structures, and the barn is reserved for grain.

One of the most characteristic and distinctive features of the barns of the Lancaster Plain is the forebay, or overshot, which projects out over the stockyard. The entire upper level extends four to six feet beyond the lower level; it overhangs the entrances to the lower level, where the workhorses and cattle are housed. These entrances have divided Dutch doors. The top half of the Dutch door can be opened to let in air and light, and the bottom half can be closed to keep in the animals and to keep out the driving rain or snow.

Originally farmers used ground-level mows (the part of a barn where hay or straw is stored) to store sheaves of grain hauled in from the fields to be threshed, but later they also used them to store hay. Typically, the upper level had bins for grain storage and a small toolroom and workshop. Straw from the threshing floor was blown out into a huge stack in the stockyard, or straw-yard, on the sheltered south or east side of the barn. Cattle were fed and manure was collected in the stockyard, which was enclosed by a stout board fence or stone wall and was sometimes even covered by a roof to protect the precious naturally occurring fertilizer. ◆

The Amish

The Pennsylvania country might not be nearly as well known today were it not for the Amish people. Twelve hundred Amish families still farm with horses rather than tractors in the area east of Lancaster. They clatter up and down the country roads in horse-drawn buggies, use oil lamps rather than electric lights, and send their children to special one-room country schoolhouses. As pacifists, the Amish live as they do to remain separate from what they see as the destructive forces of the outside world.

THE AMISH SPECIALIZE IN PRODUCTS SUCH AS MILK, BUTTER AND CHEESE, POULTRY AND EGGS, AND TOBACCO.

EVEN THOUGH THEY RAISE TOBACCO, THE AMISH FORBID THE USE OF PIPES OR CIGARETTES, BUT THEY DO PERMIT CIGARS AND CHEWING TOBACCO.

BARN-RAISING AMISH-STYLE utilizes the considerable skills of every farmer in the community.

(Left) An Amish farmer guides his enormous horse-drawn corn-planter across his fields.

THE AMISH ARE
RIGOROUS FARMERS,
RECEPTIVE TO NEW
TECHNOLOGY SO LONG AS
IT DOES NOT GO AGAINST
THEIR RELIGION. THEY
RUN SMALL ENGINES ON
GASOLINE, BATTERIES, OR
PROPANE GAS RATHER
THAN ELECTRICITY, WHICH
COMES IN BY WIRE FROM
THE OUTSIDE WORLD.
THEY HAVE SHOWN GREAT
INGENUITY IN ADAPTING
MODERN FARM MACHINES
TO HORSE-DRAWN
OPERATION.

THE GREAT VALLEY

SIMPLE WOODEN STRUCTURES hidden in isolated hollows in the steep hills of the Appalachian Uplands offer testament today to the struggles and drive of early pioneers to settle the wilderness.

The massive 200-mile-wide natural barrier of the Appalachian Uplands extends southwestward more than 1,000 miles from the Canadian border to central Alabama. Its hard hills have few areas of good farmland, with the exception of the Great Valley, one of the largest and best limestone lowlands in the country. Stretching some 10 to 20 miles wide, the Valley curves southwestward from Lake Champlain to Birmingham, Alabama. Near Harrisburg, Pennsylvania, the Great Valley begins to bend southward into Maryland and Virginia. This bend in the Valley played an important role in the settlement of the United States because it directed pioneers south.   Before the Revolutionary War, most immigrants landed at Philadelphia.

FROM AN IDEA ADAPTED FROM German farmers, log cabins became characteristic of the Appalachian Uplands and soon became the very symbol of the American frontier. (Bottom, left) Log cabin, circa 1900–1910.

(Right) The hardwood forests forced upland pioneers to learn how to scratch out an existence from one of the most inhospitable landscapes in the country. Many of their skills were gained from Native Americans with whom they shared the land.

Philadelphia provided a port for their wheat, with a frequent schedule of boats sailing to Europe. The pioneers followed the Great Valley westward across Pennsylvania and then up the Valley when it then turned southward across Maryland into northern Virginia. (In the Valley, south is up and north is down; you go up the Valley when you go south, and you go down when you go north.)

The people who lived in the isolated coves and hollows back up in the hills of the Appalachian Uplands had an existence that was quite separate from the life of the limestone lowlands of the Great Valley. These coves and hollows attracted few new people after the initial surge of settlement. They had only limited contact with the outside world, and for a century or more, time quietly passed them by.

Two types of personalities were needed to help tame the American wilderness: wild frontiersmen and stable farmers. The Scotch-Irish frontiersmen were adventurous to the point of recklessness. They were fiercely independent and roamed far and wide with ax and rifle always at the ready. They were also deeply religious. As keen observers and quick learners, they picked up their farming and farm building skills from the more stable Pennsylvania Germans. They became familiar with new crops, such as corn, wheat, and clover, and learned to grow them in regular rotations. Their crops and animals were kept under the same roof in a single building that they had learned to call a barn, as the Germans called it. Since wood had been scarce in their native countries, where cabins were built of stone or turf, the settlers learned from the Germans how to build cabins of logs cleared from the land, which became the symbol of the American frontier.

Though the rugged upland pioneers learned much from the German farmers, they learned even more from the Native Americans, who taught them how to scratch out an existence from the hard-

BALES OF HAY stacked against a shed still comprise one of the most graceful and familiar sights on the rural landscape.

(Left) Alternating rows of corn and alfalfa not only protect crops from weeds and insects but create a formal design.

wood forest without using plows or draft animals. They used hand hoes to mound up the earth into "hills," and in each hill they planted seeds of three basic crops—corn, beans, and cucurbits.

Over time, however, the cultivated patches on the hill farms were enlarged and started to look more like fields, but they remained small. Plows replaced hoes as the principal implements of cultivation. Plows required workstock—either horses or mules—and workstock required feed. Corn was the principal and sometimes only crop. It fed the people and their animals, and some of it

was distilled into whiskey. Some farmers also had a tiny patch of tobacco for cash income. These areas were increasingly commercialized and became part of the national economy.

Today, land clearance in Appalachia has become much easier. New machines such as the bulldozer, the brush hog (which can uproot and clear away fairly good-sized brush), and the bog disc (which can "ride" over obstacles and level newly cleared land) have enabled farmers to clear land with far less labor than they needed in the past, and windrows of charred, partially burned

stumps and brush are a common sight in many parts of Appalachia.

Most of the pioneers remained largely separate from the rest of the world. Up to this day, the people of Appalachia have respected the land as a source of life, and they have not expected it to be a source of wealth. They have used the land for a brief spell and then allowed it to rest until they needed to use it once again. They have been content with a small plot of cleared ground and a few livestock. Free to come and go as they pleased, to take time off whenever they felt like it, they were obligated to no

one but themselves and their Maker. But this old way of life has been fading fast since World War II, and the amount of farmland in some areas has declined.

Still, other early pioneers, the more adventurous, pushed on westward into the endless mountains of the Ridge and Valley areas, and those who found and settled the better limestone valleys were richly rewarded. They kept searching for an easy route through the mountains to the Ohio River and the western waters until they finally discovered Cumberland Gap, where Virginia meets Kentucky and Tennessee.

TODAY'S HAYRIDES may be just for fun, but when they were originally used to transport the hay for feed, bringing in a full wagon was a demanding chore, circa 1900-1920.

(Right) The verdant and fertile modern farms on the rich limestone beds in areas of the Great Valley have made local farmers prosperous.

(Overleaf) Farmers and their families celebrating the harvest in their apple orchard, circa 1894.

A FARMER CUTS OFF ripe cornstalks by hand and stacks them in bundles called shocks, circa 1900–1910.

(Left) Farms in the Great Valley are clustered like little villages. Not even the people who live there would claim that the hills and mountains of south-western Virginia are good farming country, but somehow they've managed to remain on the land and survive.

Two distinct economies and life-styles developed in the Appalachian Uplands during the nineteenth century. The farmers who quickly moved in and replaced the frontiersmen in the better farming areas of the Great Valley became prosperous. There they enlarged the cultivated patches created by the fron-tiersmen into fields and grew crops in orderly rotations. The farmers who set-tled in the more fertile limestone beds of the Valley found a much different life than their upland brothers.

There they produced all they need-ed to feed and clothe themselves, plus a surplus of cattle and horses, which could

be driven to distant markets to sell. But wheat was the early settlers' principal commercial crop. By the time of the Revolutionary War, farmers in the Valley were already growing more wheat than they could eat, and they sent a steady stream of wagons loaded with wheat, flour, dressed beef, and bacon eastward across the mountains to help feed Washington's troops. Eighty years later the Valley was the breadbasket for the Confederate Army. It continued to produce a surplus of wheat until well into the twentieth century, when it could no longer compete with large, mecha-nized producers in the West.

Farmers in the Great Valley have grown wheat in regular rotation, before hay and after corn. Wheat has been the money crop. Much of the land in the Valley has to be kept in pasture or under hay crops to protect it against erosion. Most of the hay and pastureland is used for beef cattle, which have become the primary source of farm income since World War II.

The beef-cattle business has three distinct stages—breeding, stockering, and feeding—which can be pursued on separate farms. The breeder runs a cow/calf operation: the principal source of income is the sale of calves. He keeps a herd of cows and expects each one to drop a calf each spring. An acre or two of good pasture is needed for each cow.

Breeding and stockering produce returns per acre that are modest at best,

and they require a fairly large acreage of land to produce a reasonable level of income. Most farms in the Great Valley are too small to support full-time breeding or stockering operations. Although they were enough to maintain a family and to produce a small surplus in the days of horse-drawn plows, they are undersized in the age of tractors.

Few farms in the Valley are large enough to support the people who live on them unless they have off-farm income or unless they can find some way of enlarging the size of their farm business. One way to to that is to engage in specialized farming. The northern end of the Great Valley in Virginia has two good examples: apple orchards and turkeys. Both apple orchards and turkeys require large amounts of labor and can provide full-time employment on farms that are too small for successful beef-cattle or field-crop operations.

But no matter how you look at it, not even the people who live there would claim that the hills and mountains of southwestern Virginia are good farming country. ◆

A CLUSTER OF BARNS and silos of various shapes and sizes can be seen throughout the contemporary Great Valley, signaling an active and energetic farming community.

A Personal History

JIM FRYE

AS RUGGED AS THE LAND

Jim Frye's farm was in the heart of Wear Cove which is one of the most beautiful places in the eastern United States. A limestone valley, five miles long and a mile or so wide, Wear Cove is an emerald on the northwestern flank of the Great Smoky Mountains in east Tennessee.

A fresh bearskin was drying on the wall of Frye's shed. His eyes squinted suspiciously from beneath the brim of a battered brown fedora. When I spotted the fresh bearskin, I had jumped to the romantic conclusion that Jim hunted bears to supplement his food supply, but he snorted that he would just as leave eat shoe leather as bear meat.

"Them bears eat hogs sometimes, and they get into the cornfield or the apple orchard, but otherwise they don't bother us none. We hunt 'em for fun, jes like we hunt squirrels and rabbits and coons. They's thirteen [dead] bears come offa that mountain last year. They boys round here got 'em a gang o' bear dogs, twelve of 'em. Them bears are jes as smart as you are. As soon as they hear them dogs, they head for that there [Great Smoky Mountains National] Park, 'cause they know jes as well as you do that you can't hunt 'em in there. Hit takes a good dog jes to keep up with a bear on level ground. Them dogs'll run a bear ten, twelve miles across the mountains, through the roughest places it can find. In summer a man can get et up in them mountains by snakes, but as long as you carry a light at night a snake won't bite you.

"Them truck crops I mostly raise to eat—sweet corn, beans, squash, pumpkins, potatoes, tomatoes. Corn, we feed some of it, and I've sold some. Forty years ago folks round here jes raised corn and wheat. Ragweeds would grow up in the wheat stubble, and they'd make hay out of it.

"I got a team of mares I do most of my farmin' with. I got one cow givin' milk and a cow and a heifer comin' fresh. We keep the milk cool in the springhouse. I keep four head of hogs jes for meat. I got about 75 chickens fer meat and eggs.

"Tobacco's my cash money crop. I rent four-tenths of an acre of tobacco on three different places. In January, when I got something like 1,000 pounds worked off, I take hit in and sell hit in Knoxville if the market is strong. Fore you got one crop graded off and sold, you're commencing to put out another.

"Right now I got more time'n most anything else. I'm not as gainfully occupied some times as others. I have lived in this Valley all my life—65 years if the Lord lets me live 'til September. I had twelve children, but lost one in infancy, and one was run over by a truck, right up there at the crossroads, when he was sixteen. The others are scattered from here to yonder." ◆

"I HAVE LIVED IN THIS VALLEY ALL MY LIFE—65 YEARS IF THE LORD LETS ME LIVE 'TIL SEPTEMBER."

Hay

The early settlers cleared and cultivated some remarkably steep slopes. The rocks sticking up through the turf on many a modern hillside bear mute witness to how hard people had to work the land in order to survive. They scratched the ground for crops until most of the topsoil had washed away, and then they let the hillsides grow up in pastures choked with weeds and stippled with brush and saplings. They turned their cattle out to graze on the steeper slopes and cut hay for winter feed from the gentler ones. ◈ In time a farmer might add a cluster of other small log outbuildings to the corncribs and hay barns, such as woodsheds, smokehouses, springhouses, and the like.

THE HAY AND LIVESTOCK BARNS THAT EVOLVED IN SOUTH-WESTERN VIRGINIA WERE ANCESTORS OF A TYPE OF BARN THAT BECAME COMMON THROUGHOUT THE CORN BELT.

THROUGH THE BARN'S GABLE-END OPENING THE FARMER WOULD MANIPULATE A SYSTEM OF ROPES AND PULLEYS TO LOWER A HAY FORK CLOSED AROUND A BUNDLE OF HAY ONTO WAGONS PARKED ON THE GROUND BELOW.

The hay and livestock barns that evolved in southwestern Virginia were ancestors of a type of barn that became common throughout the Corn Belt.

Farmers used to pitchfork hay into the barn through a large opening at the gable end until hay forks mounted on wheels on a track eased this backbreaking chore. By manipulating a system of ropes and pulleys, the farmer lowered the hay fork onto the wagon, closed it around a bundle of hay, and lifted the bundle to the rooftree, moved it along the track, and tripped a lever to dump the hay where he wanted it in the loft.

The principal openings in the hay barns of southwestern Virginia were in the gable ends, rather than on the sides as in the Pennsylvania wheat barn. This reflected a major change in barn function and design better suited to corn than to wheat. The upper level can be used to store hay. The opening in the gable end of the barn had a door that could be closed in bad weather. ◆

HAY IS PACKED TO THE EDGE of the opening at the gable end of the barn. The gable ends were the main openings in the hay barns of southwestern Virginia, rather than on the sides as in the Pennsylvania wheat barn.

INTERIOR OF HAY BARN, circa 1931. A few of these structures had stalls for livestock on the ground floor beneath the hayloft, but more commonly, animals and equipment were sheltered in lean-to wings, with the body of the barn filled with hay from the floor to the rafters.

(Below) The most imposing farm buildings on the hills of southwestern Virginia are the simple rectangular structures of boards that were built to store hay. Local people call them barns, but to outsiders they look like no more than large sheds.

BLUEGRASS COUNTRY

BLUEGRASS COUNTRY boasts the largest concentration of Throughbred horse farms in the country. It is also one of the most beautiful landscapes in the world.

Far and away the most famous of the limestone lowlands is the Bluegrass area of north-central Kentucky, which owes its reputation to 200 horse farms in a 15-mile semicircle north of Lexington. The Bluegrass area is a gently undulating plain underlain by limestone rich in phosphorus. The soils derived from this limestone are two to four feet deep and are remarkably fertile.

The Bluegrass horse country is the most beautiful rural area in the United States, if not the entire world. Broodmares graze placidly in neatly manicured pastures of succulent bluegrass. Dazzling white board fences follow the gentle roll and swell of the countryside. The board fences are practical as well as pretty because the high-spirited horses could easily injure themselves by running headlong into a wire fence.

Enormous old oak, walnut, and sycamore trees dot the pastures and march along country lanes that run between walls of unmortared fieldstone, now beautifully weathered and moss-grown. Trees guard the long driveways that lead back to the elegant, white-pillared mansions, and they shade the luxuriant mansion grounds. The paneled horse barns behind the mansions cost more than most suburban homes. Many homes have their own private swimming pools and tennis courts. The countryside has an air of contented wealth.

Americans first began poking into the Bluegrass area around 1766. The

early explorers were adventurous "long hunters"—Daniel Boone was among them—and land speculators, who staked out large claims. It has become fashionable to think that land speculation was bad, but in those days making a fortune on frontier land was no more wicked and sinful than making a fortune in the stock market is today. George Washington himself was not above speculating in western lands, and he might have been even more avid if he had ever laid eyes on the Bluegrass.

The Bluegrass country was the first area west of the Appalachian barrier that Americans settled and farmed. Their

WHITE-PILLARED MANSIONS shaded by venerable oak, walnut, and sycamore trees bespeak wealth and grace in the Bluegrass countryside where Thoroughbred breeding farms abound.

(Right) A Kentucky mist casts a gently haunting spell over the quiet grain and livestock farms that curl up and around the undulating limestone plain.

staging area was the Great Valley of east Tennessee, their gateway was the Cumberland Gap, and their route was the Wilderness Road blazed by Daniel Boone in 1775. The Wilderness Road actually was little more than a trail for packhorses, but it brought a steady stream of settlers from Virginia, North Carolina, and Tennessee. They drove herds of cattle, hogs, and sheep before them, and they brought horses and mules as pack and riding animals.

Kentucky had no permanent Indian settlements, but it was a prime hunting area for Shawnees from the north and Creeks from the south. Many early white settlements were abandoned when Indians attacked in an attempt to preserve their hunting grounds. The last major Indian "invasion" was ended at Blue Licks in 1782, the final battle of the Revolutionary War, but sporadic raids continued until 1794, when the battle of Fallen Timbers effectively ended the Indian influence in Ohio.

When the state of Virginia, of which Kentucky had been a part, was admitted to statehood in 1792, Kentucky was as fertile for lawyers as it was for crops. Conflicting land claims resulted from settlements that were made. Virginia had paid some of its war veterans with warrants that entitled them to 50 to 5,000 acres of land in Kentucky, depending on their rank and length of service. The warrants did not specify the location of this land, however, so overlapping

MOUNTAINEERS AND FARMERS TRADING mules and horses on "Jockey Street" near the courthouse, Compton, Wolfe County, Ky., circa 1940.

(Below) Some things just remain the same. A contemporary farmer still uses his horse to draw a plow up the hills behind his bare split-rail wooden fence.

SIGNS OF LIFE in the wilderness can still be seen in the remnants of primitive wooden houses and sheds set off in the isolated backwoods.

(Overleaf) Bluegrass farm country with its slowly rambling curves and quietude on a glorious autumn day.

claims were virtually inevitable, and the claimants with the best lawyers got the best land.

Men of wealth or influence were able to assemble large Bluegrass estates, while smaller farmers were pushed back into the poorer, hilly areas or had to move on to the newer lands that were opening in the West. Many of the large estates were acquired by well-to-do younger sons of prosperous old Virginia families. They moved in with their black slaves, transplanted their farming system and their entire lifestyle, and became the cultural and political aristocracy of the state. By 1800 they had already started to build fine mansions in

which they could enjoy the lives of country gentlemen.

From the very start, most Bluegrass planters and farmers turned west and south rather than east in search of a market for their products. They could drive herds of cattle, hogs, sheep, mules, and horses over the Wilderness Road to eastern markets, but packhorses were the chief means of transport eastward, and they could carry only small loads.

Transport downriver was much easier, and most farmers looked to New Orleans for their principal market. In the fall, when their crops had been harvested, some farmers would cut down trees, build their own flatboats, load them with

their products, and float them downstream to New Orleans for sale and shipment to Europe or to the East Coast. After they had sold their goods they broke up their flatboats, sold the wood for building material or firewood, and made the long trip home overland.

The Kentucky riverboat men were a rough lot who liked to boast that they were "half horse and half alligator," and the good people of New Orleans were only too happy to see them depart. The lucky ones could afford to buy horses, but many had to walk back to Kentucky. Their route took them northward to Natchez and then across country to Nashville by way of the famous Natchez Trace, which was proclaimed a post road as early as 1800.

The principal products of Bluegrass farms have included hemp, corn, tobacco, and livestock. Hemp (Cannabis), a tall, graceful plant, no longer a commercial or legal crop in the area, could be made into tough rope and durable duck sailcloth. The stringy fibers could also be made into burlap. In the early days no one smoked the leaves.

Despite the importance of bourbon production and tobacco farming, livestock rather than crops have dominated the Bluegrass economy from the very first. Early herds flourished, and the settlers decided that something in the grass and water gave their animals exceptionally strong bones. By 1800 large landholders were importing fine

Bluegrass's finest: Bourbon is aged in barrels from four to twelve years.

Bourbon

The very first settlers in the Bluegrass planted corn for themselves and their animals, but almost from the start they distilled part of the crop and stored it in jugs and barrels. Water from limestone springs gave them a superior product, and they discovered that it could be improved even more by storing it for a while in white oak barrels with charred interiors. The charred barrels absorbed foreign particles, mellowed the sharp taste, and gave the whiskey a pleasant amber color. By 1810 Kentucky boasted no fewer than 2,000 distilleries, and a decade later the state was shipping 200,000 gallons of whiskey a month to New Orleans alone. Bourbon whiskey took its name from Bourbon County, which originally included much of northeastern Kentucky. The county was named for the French royal family, some of whose lesser members took refuge in Kentucky after the French Revolution.

A Personal History

THOROUGHBREDS AND THOROUGHLY BEAUTIFUL

———◆❖◆———

Woodvale Farm is south of the Georgetown Paris Pike just east of the former slave village of Centerville. The main house is a story-and-a-half structure of white-painted brick. It looks small, but it has four bedrooms, and the three-car garage behind it has five bedrooms upstairs for domestic help. The house sits on a tree-shaded, eight-acre lot that overlooks the rest of the farm.

Forest Mynear, the farm manager, started working on a farm when he was 13 years old for $9 a month and board, and got what little schooling he could on the side. "I came here when I was 21 years old, and I've been here for 41 years. This has been a horse farm ever since I've been here.

"In 1948, the farm was sold to Mr. W. Alton Jones. W. Alton Jones was the youngest of seven children who were raised on a rocky, forty-acre farm near Joplin, Missouri. As an adult he went to work as a $45-a-month janitor and meter reader for a utility company, and in his spare time he took a correspondence course in bookkeeping. The utility company became the Cities Service Company, a multibillion dollar oil company, and by 1953 Jones became chairman of the board. He was one of the nation's highest paid executives. When he died, in his wallet police found a $10,000 bill, a $5,000 bill, a $1,000 bill, several $500 bills, and some small change. In his briefcase he had $45,000

IN 1983, 30 YEARLINGS WERE SOLD

FOR AN AVERAGE OF $501,495.

more, just in case he happened to want a cup of coffee or a brand new Cadillac.

"I don't see Mr. Jones but twice a year, sometimes not then. He always came for the Derby and sometimes for the fall races, too. He and Mrs. Jones would fly to Lexington in their own plane. They had a servant drive a car from New York to open the house and meet them at the airport. They threw big parties during the races."

Forest obviously ran the farm pretty much as he saw fit, and he acted as though he owned it. "We breed the mares to foal as soon as possible after the first of January because that's the official birthday of every Thoroughbred. The colts are weaned in September, and we start breaking them at the end of the year. The next fall they go for training, and they start racing when they're two-year-olds. The only race that counts is the Kentucky Derby. A few years ago one of our horses was fourth in the Derby, and another horse from this farm has also placed in it."

In 1968, Darrell Brown, an accountant from Oklahoma, bought Woodvale and transformed the preeminent farm, changing its name to Stonereath. "Our own mare, Best in Show, is worth $4 million to $5 million, but our best horse is a filly named Blush with Pride that earned $536,000 last year. We have syndicated Gold Cup, a weanling by Alydar out of Best in Show, for $2,500,000." ◆

BREEDERS WALK THEIR CHAMPION STEEDS along a paved road through lush green pastures enclosed by painted board fences.

(Left) When they are not racing around the track, Thoroughbreds relax and graze on tree-shaded acres.

breeding stock from England. They specialized in producing superior animals that could be sold as breeding stock to farmers in other parts of the United States. All of the best families were involved. The livestock shows were as much social and political as agricultural events.

Livestock were an excellent product for Bluegrass farms in the early days when overland transportation was poor because the animals could walk to market. Herds of fat cattle and hogs from the Bluegrass were driven eastward to cities on the eastern seaboard, and the spread

of cotton production south created a major market for workhorses and mules bred in the Bluegrass. Isaac Shelby, Kentucky's first governor, drove herds of mules to cotton plantations in South Carolina, and once he corralled his animals at the governor's mansion in Columbia when he spent the night there.

Beef cattle account for about 20 percent of the total income of farms in the Bluegrass, but almost from the start the breeding of horses and mules has been more important. It is not easy for a contemporary American to appreciate the numbers of horses and mules needed

before the days of the tractor, the truck, and the automobile. Farmers had to have horses and mules to work their land, city merchants needed dray horses to pull their wagons, and the ownership of fine carriage horses was a socially acceptable way of demonstrating wealth and status. Nineteenth-century Bluegrass breeders specialized in raising horses for carriages, for riding, and, of course, for racing.

Kentuckians have always had a passion for horse racing. It was a popular pastime on the frontier, and men plotted, schemed, and taxed their wits trying to find ways to breed faster horses than their neighbors. The first impromptu races

were held in the streets of towns, which provided the best available straight stretches. The streets had the additional advantages of making available a ready audience and convenient taverns for celebrations afterward. As early as 1787 the town trustees in Lexington banned horse races on Main Street, and a formal course was laid out at the edge of town in 1789. By the 1830s the Bluegrass animals had begun to compete on turf with animals from older breeding centers in the East, and these races generated such excitement that nearly half the population of Louisville turned out to watch Grey Eagle run against Wagner in 1839.

KENTUCKIANS HAVE ALWAYS had a passion for horse racing. It was a popular pastime on the frontier, and men plotted, schemed, and taxed their wits trying to find ways to breed faster horses than their neighbors. Race track, circa 1904–1908.

(Overleaf) The Bluegrass horse country, with its stately trees and its verdant pastures, boasts some of the most beautiful pastoral landscapes in the country.

It was quite proper for a gentleman to breed fast horses, but caring for them, training them, and riding them in races was considered beneath his dignity. In the nineteenth century most grooms, trainers, and jockeys were black, first slaves and later freedmen. The slaves were housed in a hamlet at the back of the estate, away from the mansion. After the Civil War the slave hamlet was turned into a freetown, and each freedman was given a house on a lot with a garden, a pigpen, and a chicken coop. The owner thus ensured himself of cheap labor. The Bluegrass area still has a number of hamlets, some of which are now inhabited by white workers rather than by blacks.

After the Civil War, horse breeding in Kentucky received an enormous boost when wealthy Northerners began flocking to the Bluegrass to buy horse farms and the status they conferred. Horse racing, after all, is "the king of sports and the sport of kings," and the fine horse grants special status to its owner.

Horse farming in the Bluegrass is a billion-dollar business. It is dominated by the breeding of fine racehorses: Thoroughbreds. The breeding, training, and racing of Thoroughbreds are separate operations. Some breeding farms train and race their own animals while some farms send their horses to specialized training facilities.

Some horse farms do manage to make money on a more or less regular basis, but a horse farm is an enormous gamble. The cost of operating even a small farm can run well over a million dollars a year, and the owner must be prepared to spend a thousand dollars as casually as most people would spend a dime. The owner may be able to write off his losses on his income taxes. As J. P. Morgan said when asked about the cost of a yacht, "If you have to ask what it costs, you can't afford it!" ◆

Tobacco

The tobacco plant has a chameleon-like ability to adapt to different environments. It is still a major cash crop in the South. In Bluegrass country, Burley tobacco is the third most important crop, and a significant cash crop. Tobacco makes so much money that even the show-place horse farms grow as much as they are allotted, five to ten acres or even more, but the tobacco field is tucked away in a remote part of the farm where it is less likely to be noticed.

A TOBACCO MARKET in Louisville, Kentucky, 1906.

(Left) The wooden barns used for curing tobacco are unmistakable with their wall of open doors that allow air to circulate within to dry the tobacco leaves. The leaves are hung to dry on wooden tobacco sticks and can be seen through every doorway.

TOBACCO WAS THE FIRST SPECIALTY CROP THAT WAS GROWN IN THE UNITED STATES. VIRGINIA FARMERS STARTED SHIPPING TOBACCO TO ENGLAND EIGHT YEARS BEFORE THE PILGRIMS LANDED AT PLYMOUTH ROCK, AND THEY HAD ALREADY IMPORTED A BOATLOAD OF SLAVES FROM AFRICA TO WORK IN THEIR TOBACCO FIELDS THE YEAR BEFORE THE MAYFLOWER ARRIVED OFF THE NEW ENGLAND COAST.

Tobacco farmers have been under fire by consumers for recent findings on the dangers of smoking, and critics have argued that they should shift to some other form of farm enterprise. Many tobacco farmers would be happy to oblige, but they lack real alternatives. Their farms are too small for livestock operations, and no other crop could produce the same return per acre as tobacco. Vegetables are often suggested, but vegetables are risky, and they require an expensive infrastructure for marketing. The tobacco-producing districts are among the few remaining islands of agriculture in the South. Ironically, terminating tobacco production would drive yet another nail in the coffin of southern agriculture and would cause great suffering for tobacco farmers. ◆

RIDING THE TRANSPLANTER through fields of early growth.

(Left) Early growth bursting into bouquets of green leaves.

(Above, right) Tobacco farmers cutting Burley tobacco, circa 1940.

(Bottom, right) Farmers impale the green tobacco leaves on inch-square wooden tobacco sticks to wilt and then hang loaded sticks on tier poles.

THE CORN BELT

The agricultural heartland of North America is the Corn Belt, a vast expanse of farmland covered with corn and soybeans that stretches for 800 miles across the rolling plains of the Middle West from central Ohio to eastern Nebraska. The seedbed and the cradle of the Corn Belt is southwestern Ohio, where three major streams of migrants converged to create the first truly American people. ◈ From the time it was first settled until after World War II, the Corn Belt of Ohio and eastern Indiana specialized in small corn-hog farms; the western Corn Belt, centered in eastern Iowa, had medium-sized farms where cattle as well as hogs were fattened; and the central Corn Belt on the Grand Prairie of eastern Illinois boasted large, cash-grain farms.

CORN, so essential to all types of farming, is one of the human wonders of the world. Moving westward in a majestic sweep across the country, the Corn Belt was cut short only when settlers encountered scanty and unreliable rainfall in Nebraska, Kansas, and the Dakotas.

The southern stream of migrants came overland through the Cumberland Gap, over the Wilderness Road, and across the Bluegrass to the Ohio River. The second stream took the river route and floated down the Ohio from its fork at Pittsburgh. When the pioneers finally reached the river that flowed west, they used flatboats—an impossible combination of log cabin, floating barnyard, and country store—piling all of their worldly goods onboard, and then floating downstream in search of a new home. The third stream of migrants, the last and probably the largest, came overland by the National Road, built in response to the needs of the newly emergent nation for routes to a marketplace for their goods. Today the National Road is paralleled by U.S. 40 and by Interstate 70 west of the Ohio River.

Cincinnati became the gateway to the Corn Belt because it allowed pioneers to leave the river easily and set out for the interior. Here settlers could equip themselves with the tools and implements they would need on the frontier; it was also their contact point with the rest of the world. When their land started to produce a surplus, they brought their products there to be processed for shipment, and purchased goods they could not produce themselves.

The settlers came to Ohio from three major areas with distinctive systems of farming. The Yankees and New Yorkers who settled northeastern Ohio had come from wheat areas. The southerners in southwestern Ohio grew corn and tobacco. The Pennsylvanians in southwestern Ohio came from regions of mixed farming. They transplanted the

THE WIDE-OPEN CORNFIELDS of Mad River Valley near Zanesfield, Ohio, circa 1913.

(Right) Early farmers enjoy the benefits of using a corn-husking machine which lightens the load and speeds the process of readying corn for human or animal consumption.

70

early farming system of southeastern Pennsylvania to the new seedbed in southwestern Ohio so successfully that the upper Miami River Valley west of Dayton even looks a little bit like Lancaster County.

A modified farming system eventually spread across the Midwest from southwestern Ohio until it reached its climatic limit in eastern Nebraska. The basic crop rotation of the Corn Belt persisted for more than a century because it was economically successful and ecologically sound. It leveled off peaks and troughs in the distribution of labor and income, it facilitated control of weeds, pests, and diseases, and it maintained and even increased the fertility of the soil. The labor requirements of the different crops dovetailed nicely to balance out the seasonal distribution of labor and to keep the farmer productively occupied for much of the year. He was spared the grief of sharp fluctuations because he did not have all of his eggs in one basket; a poor year for one crop would probably be a good year for another. This tradition remained the norm in the Corn Belt until after World War II and transformed the Midwest. It was one of the greatest human achievements ever.

The rural landscapes of the Corn Belt have been slow to change, however, and they still reflect the traditional

FIELDS OF CORN AND WHEAT prevent the growth of weeds, pests, and diseases, and keeps the soil fertile. This alternating crop system has revolutionized farming in America, changing the fates and fortunes of farmers across the country.

A Personal History

GENE MILEY

A GOOD AND FRUGAL FARMER

The land on Gene Miley's farm is tabletop flat. The fields are fenced with woven wire on steel posts topped by a strand of barbed wire. The big, white, wooden forebay barn has "Miley" spelled out in green shingles on the gray-shingled roof. The forebay and the west end of the barn have been extended by skirtlike shed roofs to make more room for hogs on the ground level. Gene uses the upper level for only hay, machinery, and general storage.

"The only money I've borrowed was when I bought this farm from my dad. I paid it off last year. I just don't like to borrow money, and I don't see the point of spending a lot of money on new machinery and equipment. We're still using a 1963 truck to haul grain and hogs, a 1965 combine, and an eight-year-old corn picker. We have three tractors—one 72- and two 52-horsepower—a corn planter, a baler, a two-year-old stalk chopper, a grinder, and the usual wagons and cultivating equipment. None of it is real new, but it all works and gets the job done."

Gene is proud to be a hog farmer. "All I need is enough crops to feed my hogs." All of the corn on his farm is picked and dried naturally, not artificially, because the hogs eat naturally dried corn better. "I pick the ears and grind my own corn and supplements. My family has lived within a mile of this farm for

"MY FAMILY HAS LIVED WITHIN A MILE OF THIS FARM FOR A LEAST 110 YEARS."

at least 110 years. Glenn, my dad, was born in 1909 and raised in a log cabin. That farm was a mile east of here. He bought this 120-acre farm in 1947. I bought him out when he retired. Carol, my wife, used to work outside, and she did everything–worked the ground, hauled corn, loaded hogs–but I haven't needed her labor since Chris, my son, started working with me."

The Miley's farm is 335 acres, more than twice the average size of farms in Preble County, according to census figures. But those figures are deceptive and are based on an official definition of a farm that is too permissive. In fact, the Mileys have one of the smallest full-time farms in Preble County, and it is successful only because Gene is such a good and frugal manager. He uses a modified form of the traditional rotation on the part of his land that is rented because his landlord requests it, but he no longer rotates crops on his own farm because the old arguments for crop rotation have lost their persuasiveness. His land is so flat that he can grow continuous corn on it with no fear of soil erosion, and he has at his disposal a whole arsenal of chemicals for maintaining soil fertility and for controlling weeds and pesky insects.

The Mileys depend on fat hogs as their primary source of farm income. This reflects the traditional importance of small corn-hog farms in Ohio and eastern Indiana. ◆

THE TYPE OF BARN that was just right for the corn-hog farming system (as pictured above and left), neither too simple nor too elaborate, was the hay barn, with sheds on either side for livestock and machinery. It was developed in the Great Valley of east Tennessee and has become the standard barn of the Corn Belt.

farming systems as well as the revolutionary changes. The buildings on many farmsteads are not well maintained, a clue that they are no longer used for farming and that the farmstead now serves primarily as a rural residence.

Most farmsteads in the upper Miami Valley have a single large barn and a variety of smaller outbuildings. The jumble of barn types reflects the diverse genealogy of the people who settled the area. New Englanders built simple frame shells two stories high, with large double doors on either side leading to a central threshing floor. These barns

originally had storage bays for grain, straw, and hay in either end, but they had no place for livestock. The animals were either housed in separate structures or they were left outdoors all winter. Later, some farmers modified these barns by building stalls in one end beneath a hay mow, but the New England barn was designed for wheat farming. It was too simple for the corn-hog farming system.

People from Pennsylvania built the familiar, two-story forebay barns with ramps leading up to the threshing floor on the upper level. These barns were too elaborate for the corn-hog farming system, which had little need for the threshing floor but did need the abundant space for hay storage.

THE BARRENS

Ten miles east of Iowa City lies the "barren" (treeless) prairie. The first white people who explored eastern Iowa found a plant cover of tall grass, knee to waist high. The English language does not have a good name for treeless grasslands, so the early settlers called them "barrens." Later they borrowed from French and called them "prairies." The prairies have enough rain to grow trees—four inches each month from April through September, with an annual total of 40 inches—and ecologists are still debating why they were not wooded. Many farmers have planted shade trees around their houses and windbreaks on the north and west sides of their farmsteads. The only natural trees are in the stream breaks along the rivers.

The silt-loam soils that developed under prairie grassland are ideal for growing corn. The decaying mat of grass roots adds an enormous amount of organic matter, or humus, to the soil, giveing it a dark brown color. The soils are derived from thick deposits of wind-blown dust, known as loess, that are three to nine feet deep. The loess, rich in plant nutrients, was deposited during the brief period after the last glacier had

Clover (left) and alfalfa (right) are perennial plants that produce new growth for several years without having to be reseeded.

Clover and Alfafa

Both clover and alfalfa, the two principal leguminous hay crops of the Corn Belt, are delicate when they are young and may not grow well if they are planted on bare ground. The farmer normally sowed them in his young wheat field. Clover was the main hay crop until a couple of generations ago because most farmers thought alfalfa would not grow well in humid areas. Corn Belt farmers almost abandoned clover once they had learned how to grow alfalfa, which produces more tons of hay per acre, and hay with a greater feed value per ton, than any other feed crop. Alfalfa puts out early growth that shades out troublesome weeds, it grows back rapidly, and it has a deep and extensive root system that helps it to withstand dry periods; like clover, it is a legume that extracts nitrogen from the air and stores it in the soil where the other crops can use it.

SMALL RECTANGULAR PATCHES of hardwood forest and can be seen as distant clumps of trees that block out the horizon in every direction.

melted, but before the seeds of grasses and other plants could migrate in from unglaciated areas to the south to revegetate the bare ground. Powerful winds swept across the exposed surface, picked up the finer particles, and swirled them eastward in giant dust storms. The wind-blown dust accumulated like driven snow, and it rounded off a countryside that was already smooth. The broad, level uplands and long slopes are gentle enough to permit the use of farm machinery, but steep enough to be susceptible to erosion when they are cultivated with corn and other row crops.

THE GRAND PRAIRIE

The Grand Prairie, as its name implies, was originally a vast expanse of monotonously level grassland. The first generation of settlers bypassed it, just as they bypassed other large prairie areas, and it was one of the last sections of the Corn Belt to be occupied. The early settlers had good and ample reasons for bypassing the large prairie areas, which were treeless. They were largely inaccessible, poorly drained, and were plagued with mosquitoes and malaria. They had tough sod that was hard to break with the plows that were available; and they were terrifying in the fall when the grass was dry, because a chance spark or a bolt of lightning could set a fearsome, wind-whipped, prairie fire that could destroy all that stood before it.

Instead, the early settlers chose easier river routes into the Midwest. Travel over land was far more difficult than travel by water before the railroad era, and the easy routes followed the stream valleys.

Few settlers ventured away from the river valleys and into the trackless prairie areas before railroads made them accessible. The railroads brought in people, lumber, fuel, farm machinery, and the more mundane necessities of life. Of even greater importance, the railroads provided a way of shipping out the goods that prairie farms produced. In the 1850s, Congress authorized the donation of large acreages of public land to private companies to pay for railroad construction.

Given the contrary environment, the settlers on the prairies of east-central Illinois had to develop a new technology. Their iron plows worked well in the forest soils farther east, but the heavy soils of the prairies stuck to the plowshares, and they did not "scour" properly. A prairie lad named John Deere made himself famous—and founded one of the nation's largest farm-machinery companies—when he began the manufacturing of "self-scouring"

THE RAILROADS THAT SERVED the Grand Prairie encouraged the development and persistence of cash-grain farming. In the early days they competed eagerly for the grain business by charging favorable freight rates for hauling corn. The trackside area in every hamlet, village, and town on the Grand Prairie is dominated by huge grain elevators where the farmers bring their crops for sale. The electric sign at the grain elevator in Watseka, Illinois, even gives the current prices of corn and soybeans instead of the usual time and temperature.

A Personal History

KENNY MATHER

CHANGING FROM LIVESTOCK FARMING TO GRAIN

◆◇◆

In 1958, many thought small farmers were doomed. But Anders Mather had kept expanding. Anders Mather was an outstanding cattleman. Each fall he bought 500 to 600 lean "feeder" cattle from ranchers in the West and fattened them to a market weight of 1,100 to 1,200 pounds. He had enough hogs to clean up the corn that the cattle wasted in the feedlot, but cattle were his first love.

His grandson, Kenny, went to the University of Iowa for a year, but he hated it. He rented the home place from his grandfather and continued to rent it when his father bought it after his grandfather died. He bought his own land but did not plan to expand any more. "I can't take any more debt," he said, "and I'm already pushing my labor supply. I'm ready to buy if I can find land I like at a reasonable price." But too much of the land is in strong hands. "Farmers bought out nonfarm owners 10 to 15 years ago when prices were good, and today most farmers want to hang on to what they've got and buy more."

Kenny grows corn and beans, and sells them as cash crops. He has no livestock. "I have more land but need less labor than my grandfather. He would probably turn over in his grave if he knew there were no cattle on the place. I stopped cattle because they weren't making any money. The development of irrigation and feed crops in the West killed

"I HAVE MORE LAND BUT NEED LESS LABOR THAN MY GRANDFATHER."

cattle feeding in this area. I had a chance to buy this farm, so I liquidated the livestock operation to get a down payment. I stopped the hogs because I had no feed storage or facilities, and it would have cost too much to build what I needed. That old corn-hog combination can't be beat year in and year out as a way to market corn. For me the ship has sailed as far as livestock are concerned, but it's a good way to market seasonal labor, and I would advise my son to get into livestock if he wants to farm.

"The hardest thing about grain farming is marketing. Your goal is to top half of the market. I spend more time at the desk than I do on a tractor, maybe 10 to 15 hours a week. We have a CPA to handle our taxes. Jan does most of the bookwork, and our data are fed into a computer at Iowa State. They send you regular reports so you can compare your operation with others like it. You can't make much money when you have to sell crops for less than it costs you to grow them. All you can do is tighten up your belt, live on capital, and hope that prices will get better.

"Things look pretty bleak right now for a young fellow just trying to get started farming, but remember that somebody's going to farm the land, even though they aren't making any money. They've just got the hope that sooner or later a good year is going to come along. But farming's not for everyone." ◆

Hogs vs. Cattle

THE BEST WAY TO EASE TRANSPORT OF BULKY CROPS WAS TO "MARCH 'EM OFF TO MARKET" ON THE HOOF AS FAT CATTLE AND HOGS.

HOGS

◧ Farmers could usually make more money per acre with hogs than with cattle because hogs have a much faster turnover rate. A cow has only one calf a year, but a sow can have a litter of six to eight pigs every six months. ◧ It takes a year to a year and a half to feed a calf to market weight. ◧ Hogs eat like hogs. They can be fattened and ready for sale in six months or so from the time they are born. ◧ Like people, hogs have shorter intestines, so they can use concentrated feed, like corn, more efficiently than cattle can. ◧ It takes four pounds of corn to make a pound of pork, but eight pounds of corn and ten pounds of hay to make a pound of beef.

CATTLE

◧ Cattle may need more land than hogs, but they can make better use of poor land. ◧ They must have roughage as part of their regular diet, but their complex digestive systems mean that they can convert roughage such as pasture, hay, and cornstalks into meat and milk. ◧ Cows require less labor than hogs.

As a general rule, smaller farms are more likely to have hogs because they have the labor and need the income, whereas large farms are more likely to have cattle. Many a hog farmer has dreamed of the day when he could "trade up" to beef cattle, which confer greater prestige.

THE ESSENTIAL ELEMENTS of the Corn Belt are all contained here: A stately steel grain bin; a simple slanted-roof, two-level wooden barn; and a wall of corn rows just about as high as "an elephant's eye."

(Overleaf) Rising like a Sphinx out of the immense stillness, the grain elevators of the Corn Belt transform its flat terrain into a powerful and dramatic landscape.

plows of steel that could be used to "bust" the tough grass sod of the prairies.

The prairies also required a new technology for making fences without the trees. The early settlers planted live hedgerows, even though the hedgerows shaded out five or six rows of crops on either side. Good farmers grubbed out their hedgerows and replaced them after barbed wire had been invented in the 1870s, but even today a few hedgerows remain as mementos of the problems of early settlement.

Settlers in many prairie areas had to drain the land before they could cultivate it. The poor drainage of many prairie areas can be blamed on the same glaciers that made the land so flat and

fertile. The great ice sheets scraped away the topsoil and gouged out the bedrock as they ground slowly southward from Canada. Once drained, the lacustrine plains became some of the most productive farming areas in the Midwest. The Grand Prairie is the best of the lot, and it may well be the finest farming area anywhere in the world.

The Grand Prairie was ripe to lead the way to the new two-year, cash-grain cropping system of corn and soybeans that has developed in the Corn Belt since World War II. In Ford County, Illinois, for example, corn has consistently occupied more than half of the harvested cropland, and soybeans have completely supplanted oats.

THE HEART OF THE HEARTLAND

The westward sweep of American agriculture, which had germinated in southeastern Pennsylvania, attained its peak in Iowa, the heart of the nation's agricultural heartland. Early settlers bypassed the Grand Prairie to settle eastern Iowa first, even though it was farther west, so farms in eastern Iowa are smaller than the cash-grain farms of Illinois but are larger than the small, corn-hog farms of Ohio and Indiana. Like the Illinois farmer, the farmer in Iowa had his hands full with crops during the summer, but like his brethren in Indiana, in the fall, when his crops were safely in the barn and in the crib, he could use them to fatten animals. With his larger farm, he could raise cattle. After the Civil War, the western Corn Belt was ideally positioned for these multiple roles. It lay between the ranching areas that were developing on the Great Plains to the west and the rapidly growing industrial cities of the East, which needed more and more beef. At the end of the Civil War, southern Texas was a veritable hive of beef cattle, but the area was a long way from any market. Cattlemen rounded up large herds of these cattle and drove them northward in great trail drives to cow towns on the railroads that were inching westward across the plains. By the time the cattle reached eastern slaughterhouses,

they were pretty tough critters, and their meat was not very palatable. Some smart and enterprising Texas cattlemen drifted on north past the expanding railroads to the lush grasslands of the northern Great Plains. There they established ranches that could produce good, lean, cattle, but these feeder cattle still had to be fattened on concentrated feed before slaughter. So feeder cattle from ranch country would stop off in Iowa to be "finished" en route to city markets farther to the east.

The family farm has today become a highly specialized business with a need for larger capital investment and a greater volume of sales than most of the stores on Main Street. ◆

THE JOHN DEERE CORN PLANTER revolutionized corn farming by allowing farmers to plant multiple rows of corn at the same time.

(Right) The grain elevators are impressive partly because the countryside is so flat that you can see across it for miles; the skyline is otherwise broken only by an occasional hedgerow.

Corn
The Corn Belt farming system was based on a rotation of corn, a small grain (either winter wheat or oats), and a leguminous hay (clover in the early days, later alfalfa). For more than a century these five crops—corn, wheat, oats, clover, and alfalfa—were virtually the only field crops grown in the Corn Belt. Corn Belt farmers used these crops primarily to fatten hogs and cattle, although they sold their wheat for cash. The small grains helped to spread the workload over the year.

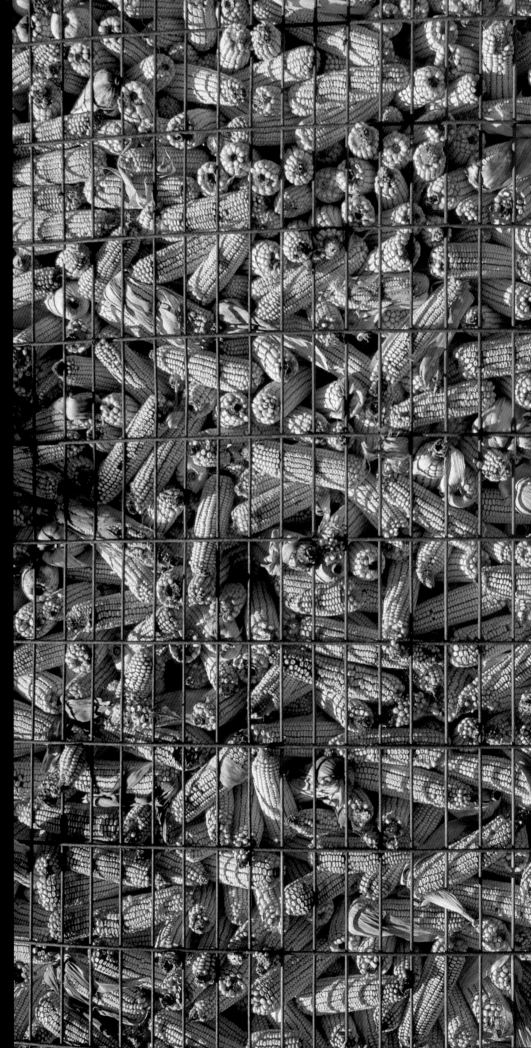

Today's Corn Belt corncrib is a direct descendent of the single-pen cribs of Appalachia. Though originally the crib could be no more than 10 to 12 feet high so farmers could heave the ears in with a shovel, today mechanical elevators enable farmers to construct larger cribs still based on the original model.

SILAGE: THE ENTIRE
CORN PLANT, STILL
GREEN, IS CHOPPED INTO
FINE PIECES AND IS
STORED IN CYLINDRICAL
SILOS. IT IS USED AS
WINTER FEED FOR
LIVESTOCK.

Corn, the linchpin of the Corn Belt farming system and the backbone of American agriculture, is a plant of subtropical origin. Unlike the small grains, corn enjoys the long growing season and abundant precipitation of the Corn Belt. Plant breeders have developed varieties of corn that have greatly extended its range, but traditionally corn has required long hot summers with scorching days and sultry nights. Corn was an excellent crop for the pioneer farmer, providing more grain per acre than wheat, and it was easily planted between the stumps on newly cleared land, whereas wheat required better soil. Corn was less susceptible to diseases and insects than wheat, but the farmer had to protect the growing crop against crows, blackbirds, squirrels, and a host of other varmints ◆

89

ALONG THE MILKY WAY

FARMER PUTTING OUT MILK cans, circa 1920. Dairy farming requires a deep commitment and is one of the most demanding lives a farmer can choose to live. It is said that two things a dairy farmer will never need are a bed and a Sunday suit.

Once upon a time American farmers tried to avoid "having all of their eggs in one basket" by doing a little bit of everything, but modern farmers, to an ever-increasing degree, have had to concentrate on doing what they do best and doing it even better. ◈ Dairy farming neatly illustrates the increasing specialization and geographic concentration of American agriculture. In 1924 almost every farm in the northeastern quadrant of the United States had a family cow or two, but by 1982 it was cheaper for many farmers to buy their milk at the supermarket than to bother with a cow. ◈ Back in 1880, when wheat was no longer a major crop in Wisconsin, Yankees, New Yorkers, and Germans, who were not ashamed of milking cows, turned the area into a dairy.

The New Yorkers introduced good dairy practices and bred better cows. They built cheese factories and developed contacts with eastern markets. The immigrants from Germany were prepared to adapt to the new country, and they were willing to do things that the natives found uncongenial: They were willing to accept the relentless drudgery of dairy farming. A dairy cow produces so much milk that the farmer must unload her every twelve hours, rain or shine, seven days a week. He must look forward to milking his cows twice a day for the rest of his life, with no thought of even a day off, much less a vacation. It is said that two things a dairy farmer will never need are a bed and a Sunday suit.

Successful dairy farming requires a long-term commitment and a love of dairy cattle. The farmer must know each cow intimately, and he must be sensitive to her whims and fancies. The herd needs so much attention and affection that he cannot trust it to a tenant, so dairy areas have the lowest rates of farm tenancy in the United States.

One of the three major dairy-farming areas in the eastern United States is sandwiched in between the Great North Woods in Wisconsin and Minnesota, and the corn and soybean fields of the Corn Belt. A second is in the productive limestone areas of southeastern Pennsylvania and Maryland near the great cities of the eastern seaboard. The third area is in the valleys and on the flanks of the Appalachian Uplands of New York and Vermont. Dairy farming produces greater returns per acre in the first two areas, but farmers in New York and Vermont rely on sales of dairy products as their main source of income.

The principal market for all three dairy-farming areas is the densely populated metropolitan and manufacturing belt of the northeastern United States. Fluid milk is bulky and perishable, and it must be produced close to those who are going to drink it. Dairy farmers who are too far from market can convert their milk into butter and cheese, which have a longer shelf life but do not fetch as good a price. Milk is seven-eighths water, and it is too expensive to carry any great distance. It turns sour very quickly, so it should be delivered to the customer within forty-eight hours of the time it leaves the cow. At one time milk was hauled to cities each night by milk

ALTHOUGH THE CARDS in the environmental deck are truly stacked against any attempt to farm at the northern edge of the Midwest—it has a short, cool growing season, complex glacial topography, a cutover forest, infertile soils, and generally poor drainage—many parts of the region provide good pastureland for dairy farming. Agricultural farm, Lansing, Michigan, circa 1910.

(Below) Dairy cows grazing in a field surrounded by a stone fence, rather than the traditional wooden one, circa 1920.

trains, which were notorious for their frequent stops. Today it moves by tank truck, but speed in getting it to the consumer remains imperative. Dairy areas have long been renowned for their good rural roads, which are essential for efficient milk collection.

Although the dairy areas are close to markets, the two northerly areas have to cope with such adverse environmental conditions as youthful glacial topography, infertile soils, and summers that are too short and too cool to produce grain crops competitively, but they are well suited to forage crops, such as alfalfa

and corn for silage. Dairy farmers harvest much of their corn for silage.

Alfalfa is a perennial crop that can be cut for hay two or three times a year. The crop yields the most tons per acre and the greatest nutrient value per ton, but it is merely the best of a long list of forage crops that includes many varieties of clover, lespedezas, vetches, and grasses, such as timothy. A farmer can pasture cattle on forage crops during the growing season, but the animals waste part of the crop by soiling or trampling it, so farmers prefer to mow, store, and then feed the crops to the cattle.

DAIRY FARMS ARE SMALLER than grain and other livestock farms; many are about the same size due to the limited number of cows that a farmer and his family can milk. Sitting through a milking can wear out even the most devoted dairy farmer, circa 1920.

A Personal History

Ralph Steiner

Dairy Farming for Generations

◆◉◆

The hard-working, German-Catholic farmers of Calumet county have big families with lots of kids. Some of the kids drift away, but many of them stay on the farm. Ralph Steiner was known to all his neighbors as a good family farmer.

Ralph's farm is on the gently rolling glacial plain four miles southwest of New Holstein, Wisconsin. "I think my grandparents on both sides came here straight from Germany. I took over the farm from my dad when he was 64 and I was 25. He was born here. Alice and I have four boys and two girls. My son Bob wants to take over the farm. I think I may let him and start working for him instead of paying him to work for me. I was 25 when I started farming, and he should have the same chance I had. I don't want to make him wait until he's 40 before he starts on his own.

"I don't buy new machines; they cost too much. I love making my own machinery. I have made a tank spreader for liquid manure, a chopper wagon, tractor loaders, and an elevator. We do most of our own building and all of our own concrete work. We built the concrete manure pond. We made our own forms of plywood and two-by-fours and had to move them 75 times when we were pouring concrete.

"I feed the cows their protein supplement when we are

"I don't buy new machines; they cost too much. I love making my own machinery."

milking them. We milk twice a day. It takes about an hour and a half in the morning, but only an hour and a quarter in the evening because the time between milkings is shorter. I start milking the cows when they are two years old, and four 305-day lactations is pretty good. After that, the cows go to make hamburger. I average about 16,000 pounds of milk per lactation." Ralph has had cartilage operations on both knees from sitting in the milking position, and Alice usually has to help him out, but this evening his daughters were on duty. They both knew exactly what they were doing.

The girls opened the door to the loafing yard, and the cows came jostling in and found stanchions. A big bucket of pellets was placed in front of each cow, and each cow's udder was then washed. Their teats were attached to the milking cups and coupled to the overhead vacuum line. A wheelbarrow with high-moisture corn was wheeled in, and two scoops were placed in front of each cow, three in front of the high producers.

Ralph kept moving cows from one side of the barn to the other because only one side has a milking line. One old cow—"a problem cow," Ralph muttered—kept pawing at the milking cups with her hoof, trying to scrape them off. Ralph had to sit on an overturned bucket beside her while she was being milked to make sure she behaved herself. ◆

Successful haymaking requires dry weather. The plants must be dried in the field for several days after they have been mowed, until their moisture content is less than 20 percent. A sudden shower can ruin a good crop of hay. Modern glass-lined, blue, metal silos enable farmers to harvest and store forage crops when their moisture content is high. The metal silos are gastight, and the plants in the silo remain in the same condition as they were when they were harvested. This moist product is called haylage. The new metal silos are expensive, but they permit the farmer to harvest forage crops when they are at the ideal stage of maturity, and he does not have to wait for good haymaking weather.

Forage crops are the most economical source of nutrients for animals, but they are bulky in relation to their value, and they normally are fed to animals on the farms where they have been grown. Areas that have an abundance of forage crops usually have cattle of some kind because cattle and other ruminants have marvelously complex digestive systems that enable them to digest roughage. They can digest fibrous plants, such as grasses, and they can digest the fibrous parts of plants, such as their stalks, stems, and leaves. Poultry, pigs, and other animals are pretty much restricted to eating the seeds and fruits of plants.

Most dairy farmsteads are well maintained and attractive. The dairy

Preserving Milk
The women of the farm preserved the surplus milk of summer by making it into butter and cheese. They made cheese by adding rennet—the membrane of a calf's stomach—to sour milk to make the curds coagulate. They pressed, molded, and cured the curds, and fed the watery whey to hogs. They skimmed the cream from whole milk to make butter and fed the skim milk to hogs. They put the cream in a churn and agitated it until the butter formed and could be worked out of the buttermilk with a wooden paddle. They salted the butter heavily to keep it fresh and packed it in wooden kegs. Some people fed the buttermilk to hogs, but others drank it themselves.

The farm wife took surplus butter and cheese, and anything else her family did not need, such as eggs, salt pork, lard, and smoked meats, to the country store, where she bartered them for groceries. The butter often spoiled, even though it was heavily salted, and people complained that it was unfit for human consumption and could be used only for grease.

THIS VERMONT DAIRY FARM, as well as the one pictured on the following page (overleaf), are characteristically charming examples of the unique character of the dairy farm. Dairy farms are among the most beautiful farmsteads in the country. Since dairy farmers spend most of their time on the farm they are immaculately maintained and lovingly designed.

farmer spends most of his time on the farm and likes to keep it looking nice. Dairy farms are smaller than grain and other livestock farms, and many are about the same size because there is a limit to the number of cows that a farmer and his family can milk. Small farms beget a dense rural farm population and prosperous market centers, which are closely spaced because dairy farmers cannot remain away from their cows for more than a few hours at a time. The market centers have cooperative creameries where butter and cheese are made.

The tradition of cooperation to achieve common goals has carried over into the political arena, where dairy groups have been unusually effective.

The College of Agriculture at the University of Wisconsin played a prominent role in developing dairy farming in the state during the latter part of the nineteenth century. The agricultural scientists at the university encouraged dairy farmers to switch from dry fodder corn to silage for winter feed, which would enable them to breed their cows to produce milk throughout the year. In

THERE'S NOTHING QUITE as evocative as the sight of of an old red tractor standing in a flimsy wooden barn.

(Left) The contemporary division of cropland into sections of soybeans and corn has replaced oats and other grains and has become the favored combination among farmers. Nearly half of the country's farm income over the last two decades has been in the sale of corn and soybeans.

1900, Wisconsin had fewer than a thousand silos, but by 1914 nearly every dairy farm in the state had at least one, and many farms had several.

The first silos were built of wooden staves, but soon concrete became common, and cylindrical, concrete silos became the hallmarks of the dairy landscape.

Cylindrical, concrete silos tower over red barns with huge haylofts above solid masonry ground floors. Numerous windows admit light and air to the lower level, where the cows are milked. Larger farms have several silos.

THE DAIRY-FARMING REGIONS

The Milky Way that once extended from Minnesota to Maine coagulated into scattered clusters in response to different environmental constraints and different economic opportunities. These differences are highlighted by dairy farms in Calumet County, Wisconsin; in Madison County, New York; and in Hampshire County, Massachusetts.

Calumet County is east of Lake Winnebago in the dairy country of

northeastern Wisconsin. The rolling plains are the gift of the glaciers, which plastered the countryside with deep deposits of lime-rich debris when they melted. The soil is sweet and fertile, and most of the land is gentle enough for easy cultivation.

Wisconsin dairy farms have continued to improve. The availability of the new short-season corn varieties has enabled some dairy farmers to switch to cash-grain farming in the warmer and flatter parts of southeastern Wisconsin, where they can commute to city jobs and grow grain on the side, and on the flat prairie of southwestern Minnesota, where dairy farming has always been somewhat marginal.

Improvements in the technology of making and storing forage crops after World War II enabled dairy farmers to bring their pastures and hay crops to their cows, which is more efficient than turning the cows out to graze. He could keep the animals in a small, enclosed feedlot near the barn and place their feed in a trough on one side of the feedlot.

The improved, and more expensive, technology has forced dairy farms to become larger and more productive. Better breeding, better feeding, and better management increased the average milk yield per cow from 6,000 pounds in a 305-day lactation period in 1940 to 12,300 pounds in 1980 for the same

SHADED BY A SLOPING ROOF, dairy cows are standing at stanchions, where high-moisture corn and haylage are served up before them.

A Personal History

ED MONTAGUE

TOO IN LOVE TO LEAVE

Ed Montague was milking 25 cows and just barely managing to hang on by his fingernails. He owned 350 acres, but he had to buy much of his feed because only 31 acres were tillable. The rest of his farm was severely cutover hardwood forest or steep and stony pasture that, in his words, was "hardly worth anything. Juniper has ruined all of our pastures, and you can't kill it."

Ed's farm was in the hills west of the broad valley of the Connecticut River. Roger Harrington, the ebullient county agent in Hampshire County said, "The hills and the valley are like night and day. The hills are mostly woodland and stony pasture, but parts of the valley are pretty good farming country. Valley agriculture is mostly specialty crops such as vegetables, potatoes, and tobacco. The dominant stock in the agricultural section of the valley is Polish and Czech. The Yankees can't take the backbreaking work like the Poles, and they've been pushed back up out of the valley and into the hills. The Yankee hill areas are still pretty much dairy country—small farms that are struggling to get by."

Ed's farm had a handsome, old two-story, white frame farmhouse that sat next to the road. A large unpainted shed connected it to a weatherbeaten old hay barn at the back. "My mother's people built this house and the old barn in 1813. We

"A MAN WOULD HAVE TO BE CRAZY TO TRY TO FARM HERE. WE ONLY MAKE ENOUGH TO KEEP THE WOLF FROM THE DOOR."

installed plumbing in the house in 1913. We cook with oil, but we still burn wood in winter.

"About all the woods are good for is maple syrup. We cut them hard during World War II. It's got to grow a lot faster than it does now before it'll be any good. There won't be any more cut as long as I live because it's not worth it. I tap about 600 buckets of maple sap every March and get around 100 gallons of syrup. People come right up to the house and want to buy maple sugar when they see smoke coming from the sugaring house, but most of it is already ordered in advance for $5 a gallon.

"I do day work a good deal. I get a little roadwork every now and then. I used to work in the woods full-time and did chores morning and evening. I still do a bit of sawing. Every Saturday night in the fall I run hayrides for 40 people, mostly old Jewish ladies from New York, who come up to a hotel in Northampton when the leaves are changing color. I put bales of hay in the wagon for them to sit on, hitch up the horses, and keep them out for about two hours.

"A man would have to be crazy to try to farm here. We don't make enough to pay our bills, only just enough to keep the wolf from the door. I only do it because it's the old home place. If I didn't love it, I'd leave. If any of my boys wanted to buy the farm, I'd sell it tomorrow." ◆

period. The best dairy cows produce more than 20,000 pounds of milk.

Wisconsin dairy farmers have consistently produced more milk than people could drink. A relatively small area near a city could produce all the milk its people could use, and milk produced in areas farther away was hauled to creameries and made into butter and cheese. In 1959, nearly every crossroads in Wisconsin seemed to have a creamery on one corner and a tavern across the road. Creameries in the more distant areas made butter because it is the most concentrated product and can bear the greatest transport costs. A pound of butter takes twice as much milk as a pound of cheese. Dairy farmers took skim milk and whey from the creamery back to their farms and used it to fatten hogs. Many also supplemented their income by raising cash crops, such as sweet corn, snap beans, green peas, or other vegetables that had little direct relationship to their dairy operations.

Cows produce the most milk in the spring and early summer when they are grazing on the new green growth, but people drink the most milk in the fall and winter, when production is lowest. Urban milksheds used to expand in the fall and winter, when supplies were low and the price of milk was high, but in spring and summer, when dairy farmers produced lots of milk, they could hardly give it away.

Fruits and vegetables At one

time most farmers and even many small-town people grew much of their own fruits and vegetables. Eventually farmers realized that there was money to be made by producing top-quality produce as field crops rather than as garden crops. As a result, the production of vegetables and fruits, even more than dairy farming, has become increasingly concentrated geographically. The requisites for a good fruit-producing district are abundantly well satisfied in southwestern Michigan. The area is close to Chicago and Milwaukee, it has the infrastructure necessary for producing and marketing fruit successfully, and it enjoys favorable growing conditions. Due to greater awareness of the satisfactions of eating fresh fruits and vegetables, people insist on buying local produce, resulting in the burgeoning of indoor and outdoor farmers' markets along the countryside.

VICTORIAN MANSION in upstate New York. Farmers in upstate New York made fortunes growing wheat between 1800 and 1850, and as a result, handsome Victorian mansions like the one pictured above came to line the streets.

UPSTATE NEW YORK

Upstate New York, much like eastern Wisconsin, was remodeled by glaciers, but in New York the glaciers were bulldozing their way uphill into the uplands of Appalachia. They scraped smooth the hillsides and gouged out the valleys, some of which later filled with water to become the Finger Lakes. The ice, when it melted, plastered the countryside with stony debris, and it left a legacy of thin, sour soils that require careful management.

Upstate New York is downwind from the Great Lakes, which cool and humidify the air masses that cross them. Gray overcast days are common and precipitation is abundant, even excessive, but it is the kind of gentle steady rain that soaks the soil without eroding it seriously. Spring comes a week or two later in New York than it does in Wisconsin, the growing season is several weeks shorter, and the summers are a few degrees cooler. These slight differences near the margin can be critical. It is possible to grow corn in upstate New York, but hay is a safer and more reliable crop. Part of the farmland is wooded, and part of the cleared land is used for hay and pasture.

Upstate New York was at the forefront of American agriculture between 1800 and 1850. Farmers made fortunes growing wheat and could afford to experiment with new ideas. The area became a leading center of agricultural innovation, and prosperous farmers built fine mansions for themselves. The handsome small towns and pleasant rural areas of upstate New York have been discovered by affluent people from New York, Buffalo, Rochester, Albany, and other large cities within easy driving distance.

After 1850, the center of wheat farming shifted to the Midwest, and most farmers in upstate New York shifted to dairying. ◆

Dairy vs. Beef Cows

Dairy cows require better nutrition than beef cows if they are to produce copious quantities of milk, and the cool, moist areas where good forage crops can be produced are the domain of dairy cattle. The best breeds were developed in the mild maritime areas of northwestern Europe, and dairy cows are happiest where the summers are not too hot.

ONE POUND OF GRAIN CORN SUPPLIES THE SAME ENERGY AS THREE POUNDS OF ALFALFA HAY, ALFALFA HAY HAS NEARLY DOUBLE THE PROTEIN CONTENT OF CORN SILAGE ON A DRY-MATTER BASIS, AND A POUND OF SILAGE IS THE DRY-MATTER EQUIVALENT OF ONLY A THIRD OF A POUND OF HAY.

Cattle are common wherever environmental conditions inhibit crop production—where the slopes are too steep, where the rainfall is too scanty, or where the growing season is too short and too cool for crops to mature and produce fruits and seeds. One of the best ways to use steep slopes and subhumid areas is to grow grass for cattle to graze.

Dairy Cow

A dairy cow is a far more complicated piece of business than many people seem to realize, and dairy farming requires much more than simply shoving hay and silage in at one end of the cow, and pulling milk out of the other, and trying to dodge the inevitable byproduct of the feeding operation.

A cow needs a basic maintenance ration of two or three pounds of hay (or its dry-matter equivalent) each day for every hundred pounds of body weight. At least a third of a cow's daily ration, by dry-matter weight, must consist of roughage.

A dairy cow can produce up to 70 percent of her milk potential if she is fed nothing else, but she needs concentrated feed if she is to fulfill her potential for milk production. She simply cannot consume enough roughage to produce milk at her maximum ability.

The dairy farmer must supplement the roughage in the cow's diet with more concentrated sources of energy, such as grains and other feeds that are rich in proteins, minerals, and vitamins. A dairy cow normally gets a pound of grain for each three pounds of milk she produces. The farmer begins by figuring how much of her energy and other needs she is getting from the roughages, and then he adds the most economical concentrate that will keep her producing milk with the greatest efficiency.

The dairy farmer knows from experience how much hay or silage his cows should get, and he does not need to weigh each cow to find out. He carefully records each cow's milk production, and he cuts back on her concentrates when her production starts to drop off because the feed she does not need to produce milk will wind up on her body as undesirable surplus flesh.

Cattle

Cattle will cheerfully eat many different things, and the kinds of forage vary so greatly from one area to another that specialists in animal nutrition use the concept of "energy" as a yardstick for comparing the values of different feeding stuffs. They have developed some complex and highly sophisticated techniques for measuring energy, which is a bit like the way calories are used.

Cattle can satisfy part of their daily energy requirements by eating roughage, such as pasture, hay, haylage, and silage, and they must eat roughage to keep their complex digestive systems in proper working order.

Cattle can live on roughage.

The beef farmer gives his cattle as much grain and concentrates as they will eat in order to put flesh on them as rapidly as possible. ◆

A cow has a gestation period of 283 days, and usually she is bred again 60 to 90 days after she has had a calf. Farmers used to breed their cows to calve in late spring to take full advantage of lush summer pastures,

but this led to a glut of milk in summer and a shortage in winter. The modern dairy farmer breeds enough cows each month to ensure a steady supply of milk throughout the year because he has learned how to store summer plant growth for winter feed. Two-thirds of the dairy cattle in the United States are bred by artificial insemination, which has improved the quality of the dairy herds because frozen semen enables farmers to use better sires than they could afford to purchase in the past. The use of artificial insemination has also made dairy farms safer places because dairy bulls have a well-deserved reputation for being mean and dangerous.

They Used to Call It the Cotton Belt

A FIELD OF MATURE COTTON. Though the traditional Cotton Belt no longer rules southern farming, cotton has a modern kingdom that extends its reach across the U.S., courtesy of mechanized production.

For more than a century and a half, cotton was the single crop that dominated the flat to gently rolling landscape of the South. For many people the South is still synonymous with the traditional Cotton Belt: a region that at one time stretched across the South from New Mexico to North Carolina and was cultivated by small family farmers. In 1950 the U.S. Department of Agriculture published a still widely-used map showing that same area as the Cotton Belt. The facts are otherwise. A whole host of problems, including the 1921 invasion of the boll weevil, which destroyed more than 2 million bales of cotton each year, gradually whittled away cotton's domain, and the traditional Cotton Belt ceased to exist.

Today, the growing of cotton is an entirely different business: it is a vast commercial industry reaching straight across the United States from Virginia to California. In 1820, however, most of the crop was grown on the Piedmont (the foot of the mountains) of South Carolina and Georgia, and the frontier of white settlement was no farther west than the present site of Atlanta. By 1840 cotton production had moved westward to the limestone valleys of northern Alabama, the Black Belt of Alabama and Mississippi, the Red River Valley of Louisiana, and, most notably, to the brown loam uplands north and south of Natchez and Vicksburg in southwestern Mississippi. The Blackland Prairie,

which is at the margin of the area where cotton can be grown without irrigation, was settled after the Civil War by people whose small farms worked with family labor and turned the Blackland into a major cotton-producing district.

The contemporary southern land-scape still retains remnants of the old cotton farms. The early cotton farmers' terraces still snake across green pastures through wooded areas and across newly plowed fields. They are a splendid reminder of the vast amount of land that was cultivated and exhausted in years past by small farmers living off the land. Numerous abandoned small farmhouses give way to farms that have been enlarged. Working farmsteads and

EARLY COTTON PICKING required tremendous stamina. Black workers labored under the hot August sun to fill wagons headed for the cotton gin, circa 1915. After emancipation and during World War II, the mass exodus of black and white share croppers left farmers without sufficient labor to work the fields, hastening cotton's decline.

implement dealers still find space for cotton-picking machines and wagons, but combines, grain drills, and metal grain bins are elbowing them aside. The cotton gins in the small towns look old, tired, and rusty, but the grain elevators are spanking new. The area that was called the southern Cotton Belt is a historical relic.

The South is different from the rest of the country. For one thing, the nature of farming in the region still provides an odd mixture of commercial and "way of life" farming. It is also a tough place to make a living by farming. Long, hot summers drenched by heavy precipitation provide ideal growing conditions for most plants, but there is a trade-off between climate and soil; the areas that have the most genial climates often have the worst soils.

The South is one of the wetter parts of the world, but its rainfall is unpredictable. Farmers normally expect two or three weeks without rain, but in bad years severe droughts can last all summer long. With a few notable exceptions, the soils of the South were no better than mediocre to begin with, and the cultivation of row crops allowed torrential rains to strip away the topsoil and carve gullies deep into the subsoil.

In the early seventeenth century, before cotton farming, southern farms in eastern Virginia and Maryland were usually tobacco plantations. The English who settled there discovered that they could not transplant traditional English field agriculture to the semitropical environment and turned to tobacco as a cash crop. By 1690, planters were also growing rice on reclaimed swamps. After 1740, the plantation model was transferred to the cultivation of indigo, which was fermented to extract a deep blue dye. Indigo was an unpleasant and unhealthy crop that was grown by slaves on large upland landholdings inland from the rice plantations. It enjoyed a brief period of prosperity after the Revolutionary War, but overproduction in other areas depressed the price so severely that most indigo farmers had switched to cotton by 1800.

Cotton was little more than a curiosity in 1790 because separating the seeds from the tenacious fiber was so slow and expensive. But later, in 1793, after the development of the cotton gin (short for engine), which did the work fifty times faster, production doubled each decade thereafter until the Civil War. Though it was easy to grow, cotton demanded so much labor that even a large family could handle no more than 10 to 15 acres. The backbreaking tasks of chopping and picking required little skill but tremendous stamina. The

semitropical climate that was great for cotton was also great for a whole host of weeds, and all summer long people spent their days in the broiling sun chopping out the weeds with hand hoes. Picking started in August and lasted for a couple of months. Schools closed for a cotton-picking "vacation" so the children could help. The pickers had to bend from the waist to pick the ripened bolls because the cotton plant is a bush that grows only thigh high.

Most of that work was done by slaves. After the Civil War, however,

THE BILLUPS PLANTATION in Greenwood, Mississippi. The end of plantation life began with the freeing of the slaves and, later, with the mass migration of black and white workers to northern cities for jobs in World War II war munitions plants.

(Second overleaf) Dust forms a smoky haze over fields of both unpicked and harvested cotton.

A Personal History

DON GRANT STILL HANGING

ONTO COTTON IN SOUTH CAROLINA

◆

The D. D. Grant and Son store is a neat, white frame building that dominates Cypress Cross-roads, South Carolina. Two modern gas pumps stand in front of the store, the porch shades a well-worn wooden bench, and the windows have metal bars to discourage attempts at breaking and entering. A hundred yards down the road is a cotton gin, a great, hulking, nondescript box of rust-blackened corrugated metal, with shed roofs tilted out over the unloading areas.

Don Grant, 56, is one of the few remaining cotton farmers in South Carolina. In 1949 nearly 100,000 farmers in the state grew cotton, but in 1982 they numbered fewer than 500, and Don alone grew more than 1 percent of the total acreage of cotton grown in the entire state. "I'm farming 2,000 acres," he said, "half in cotton and half in soybeans. We own about 60 percent and lease the rest. My mother's father farmed here.

"In 1943 this farm had one tractor with a disc harrow, and everything else was mules. I couldn't even begin to tell you how many tenant houses there were. Seems like every time you turned a corner there'd be a house with a family in it. They cost me money for taxes and upkeep, and I can't afford to fix them up as nice as I would like them to be.

"Seven men work on the farm, and two help around the

"COTTON IS OUR BREAD AND BUTTER. I JUST LOVE COTTON. IT'S BEEN GOOD TO US."

store and the gin." Every one of them has worked for Don for at least 15 years. "They are all my friends," said Don, "but we don't socialize." They were a proud and loyal lot. Each one wore clean and neatly pressed coveralls or a light blue shirt with his first name sewed right onto the shirt. Any filling station in the country would be happy to have such a trim and tidy crew.

"Cotton is our bread and butter. You need at least 600 pounds to make any money. I wouldn't want to do without soybeans, but I just love cotton. It's been good to us. The challenge is growing the crop, and it's just not as much fun once it's been picked. My grandfather and father bought an existing gin in the 1930s, and we have improved and modernized it through the years. Although I could do without it because there's another one half a mile down the road. We sell our cotton through cotton merchants. We need to pay more attention to marketing, as much as we pay to production, because a cent a pound can make a lot of difference when you sell 1,500 500-pound bales.

"Soybeans came in during the 1950s. They were a good money crop, simple and not too expensive to grow, low labor cost, and readily marketable. They replaced the corn and oats we had grown to feed the mules. Livestock? I don't want anything on the place that I have to feed. I'm a cotton farmer." ◆

WORKERS UNLOAD COTTON from horse-drawn wagons into the cotton gin, where seeds were separated from the fiber at fifty times the rate of doing so by hand, circa 1890.

(Left) Day laborers picking cotton near Clarksdale, Miss., circa 1939.

Soybean pods. Soybeans have replaced cotton as the leading crop of the region and have supplanted many crops across the country.

Soybeans

Since World War II the acreage of soybeans in what was formerly the southern Cotton Belt has been increasing at the expense of cotton and corn. Soybean production is concentrated in level areas where the land is suitable for large fields and modern machinery because they produce only a modest return per acre, and a farmer must plant a substantial acreage of soybeans in order to generate a reasonable return. Initially farmers grew soybeans as a fill-in crop when government programs restricted their acreage of other crops. Soybeans still remain a secondary crop in the minds of many farmers. The soybean crop has made money when other crops have not, but few farmers have developed the affection for soybeans that they have for cotton, and few boast about being soybean farmers. The traditional land of cotton has become the land of double-crop soybeans and winter wheat.

the emancipation of the slaves forced planters to reorganize their operations. Though the planter still owned the land, he had no one to work it; while the freed slaves had no land at all. This dilemma was resolved by the infamous system known as sharecropping, whereby former slaveowners would lease parcels of land to be worked by sharecroppers at a price that would leave them with little of their own livelihood.

Cotton remained king in the South from the Civil War until World War II, but a whole host of problems gradually whittled away its traditional domain. One of the mightiest predators was the boll weevil, which sneaked into southern Texas from Mexico in 1894. By 1921 they had invaded the entire South, destroying more than 2 million bales of cotton a year. Eventually, farmers learned how to control boll weevils, but much of the country east of the Mississippi River never fully recovered from their depredations. Soil depletion and erosion, discouraging prices, acreage-control programs, competition from other areas and from synthetic fibers, all had buffeted cotton production in the South, but its death knell was not sounded until World War II, when cotton planters lost their labor supply. The war prompted an exodus of black people to northern cities, and many white workers also left the land for jobs in the new war plants that were being built in towns and cities in the

THESE MESH MOTH TRAPS provide a protective net for cotton growing in the fields beside it and also add a dramatic visual element to this landscape at sunset.

(Left) Rows of young cotton in the early morning light provide a picture of beauty, made even more beautiful by the purely functional nature of the design.

South. The loss of labor led farmers to mechanize, but they simply couldn't keep up. The new technology led small farmers on marginal land to simply stop trying to grow cotton.

But today cotton has returned, not to the traditional Cotton Belt, but as a large farming industry that reaches across a larger landscape. With mechanization, today cotton covers more than 12 million acres of the country, producing approximately 14.5 million bales as a result of better land use, improved plant varieties, fertilization, irrigation, and better control of disease, weeds, and insects. ◆

Cotton

Fall is harvest time for cotton, as it is for most plants. After being harvested, the stalks are cut down and turned under the soil. In the spring the land is plowed again, and the soil is broken up and formed into rows into which seeds are dropped, covered, and packed with earth. Weeds and grass—which compete with the cotton plant for soil nutrients, sunlight, and water—are removed by mechanical cultivators.

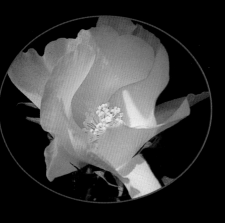

COTTON'S EARLIEST ORIGINS

NO ONE KNOWS HOW OLD COTTON IS. BITS OF COTTON BOLLS FOUND IN MEXICO WERE PROVEN TO BE AT LEAST 7,000 YEARS OLD.

3,000 YEARS BEFORE THE BIRTH OF CHRIST, NATIVES OF EGYPT'S NILE VALLEY WERE MAKING AND WEARING COTTON CLOTHING.

ARAB MERCHANTS BROUGHT COTTON CLOTH TO EUROPE ABOUT 800 A.D.

Approximately two months after planting, flower buds called squares appear on the cotton plants. In another three weeks, the blossoms open. Their petals burst out in colors that range from creamy white to yellow, pink, and finally dark red. After three days, they wither and fall, leaving green pods shaped like tiny footballs, called cotton bolls.

Inside the boll, moist fibers grow and push out from newly formed seeds. As the boll ripens, it turns brown. The fibers continue to expand under the warm sun. Finally, they split the boll apart, and the fluffy cotton bursts forth into what looks remarkably like white cotton candy. At this time the crop is harvested. Today, harvesting is done by machines that gather cotton fifty times faster—and is a lot less backbreaking—than when workers used to pick it by hand. ◆

(Clockwise from top) Yellow cotton bloom formed two months after planting; the bloom darkens into pink; after petals wither, a green pod called a cotton boll forms; finally the pod bursts open into a cotton boll.

(Preceding page) Cotton is sprayed to protect crops from weeds and other predators.

(Left) Red cotton pickers.

(Left bottom, left) Boll buggy dumping cotton into module maker.

(Left bottom, right) Two modules of cotton from Farmers' Coop Gin.

SOUTHERN SPECIALTIES

YOUNG GIRL AT PEANUT HARVEST TIME, circa 1900. Farmers in the six states of the Old South grew millions of acres of cotton in the 1940s, but today that land is used for peanuts, sugarcane, rice, and pine forests, which are scattered like polka dots across the map of the South.

he South has been a splendid laboratory for studying the birth and death of agricultural regions. Traditionally, the South had been a one-crop farming area, but cotton did not hold undisputed sway over the entire South even in the days when it was king. Tobacco, peanuts, rice, and other specialty crops have had their own smaller fiefdoms. ◈ Specialty crops are cash crops that fetch good prices and dominate the economies of the areas in which they have been grown. Their geographic concentrations originated by historic accident when someone started planting the crop in an area that was environmentally suitable. A far larger area is suitable for growing them, but their acreage is limited by the meager amounts the public consumes.

The specialty-crop areas of the South are small islands of cultivated land in a sea of dark pine forest, changing only slightly over the years. These crops are grown in small patches on the best land within the area. These islands of specialty-crop production are scattered like polka dots across the map of the South. The margins of these areas have contracted because productivity has outstripped demand. Breakthroughs in technology doubled the yields per acre in the two decades after World War II. The combination of rising yields and slack demand has forced some marginal areas to reduce or even cease their production and abandon the land.

The broad outlines of the principal tobacco, peanut, rice, and sugarcane areas were formalized by agricultural legislation in the 1930s. Subsequent agricultural legislation has continued the allotment system of acreage controls and fossilized the geographic pattern of the 1930s. It has been extraordinarily difficult for an individual farmer to break into the system and start growing a crop if he does not already have an allotment, and it is virtually impossible to start producing it in a virgin area.

POULTRY

Poultry—broilers—have been the true livestock success story of the South since World War II. Specialized poultry production began in the 1930s in "problem farm areas," areas that had large numbers of small farms with a low income per farm and per family. In 1949, farmers in the six states of the old cotton South sold only 95 million broilers, but in 1982 they sold more than half of the nation's meat-type chickens. They have changed chicken from a luxury to a staple in the American diet.

EARLY FARMS (top left and right), always had their family of chickens. Today, larger-scale poultry farming has become an attractive option for many small farmers, especially those who can find off-farm jobs, because they produce regular income and require only a few hours of work each day.

Once nearly every farm had a small barnyard flock that survived by scavenging, and chicken was reserved for Sunday dinner, but the broiler business has become so highly specialized and efficient that chicken is our cheapest meat.

Chickens convert feed more efficiently than cattle or hogs. In 1980 it took eight pounds of corn to produce a pound of beef, four to produce a pound of pork, and only two to produce a pound of chicken. The price of chicken has increased only slightly since World War II, while the prices of beef and pork have more than tripled.

The broiler-producing district of northwestern Arkansas is located in the Springfield Plateau section of the Ozark Uplands. Headward erosion by streams has fretted deep valleys in the edges, but the central strip is a wide, gently rolling upland. The soil is too thin for successful crop cultivation, and most of the land carries verdant pastures that are grazed by beef and dairy cattle.

PEANUTS

In 1919 the good citizens of Enterprise, Alabama, erected a monument to the boll weevil because it had brought them new prosperity when it forced local farmers to switch from cotton to peanuts. "The world's only monument to a pest" graces a fountain at the town's principal intersection.

The area east of Enterprise along the Alabama-Georgia line has become the nation's leading peanut-producing district. This area anchors the southwestern end of the Inner Coastal Plain, the second most important agricultural area in the South. The northeastern end of the Inner Coastal Plain is anchored by the other major peanut district, along the Virginia-Carolina line. Cotton was grown on the Inner Coastal Plain, but the sandy loam soils are a bit droughty for cotton. The level topography is well suited to large fields and large machines, however, and this area really came into its own as an agricultural region after World War II.

Peanuts were brought to the United States from Africa on slave ships. They were grown widely in the South before the Civil War. Yankee soldiers developed a taste for them during the Virginia campaigns, and the Virginia-Carolina area began to specialize in peanut production to satisfy the new national demand. Farmers in the Georgia-Alabama area turned to peanuts after the boll weevil wiped out their cotton crops around the time of World War I. Farmers were urged to grow more peanuts during both world wars because

they are a good source of the vegetable oil that cotton had previously yielded.

Farmers planted alternating rows of corn and peanuts and turned hogs into the field when the crops were ripe. Some people say that a real Virginia ham can come only from a hog that has been fattened on peanuts.

At harvest time in September and October, farmers dig up the peanut plants, shake the dirt from the nuts, let the vines wilt on the ground for a few hours, and then stack them to dry on

PEANUTS ARE NOT NUTS, despite their name. They are legumes, close relatives of beans and peas. They grow on bushy plants only a foot or so tall. The stems droop to the ground after its flowers have bloomed, and the nuts mature in shells beneath the soil.

(Right) Large tractors speed up the harvesting of peanut crops.

poles eight feet high. The slender stacks, no more than three feet thick, used to stand in the fields for four to six weeks until the pods were dry enough to be threshed from the vines. But the harvest has been mechanized since World War II. Special plows dig, shake, and windrow the vines. The vines are allowed to dry for a few days, and then the pods are removed by combines that deliver them to metal wagons in which they are dried by hot air.

About one-third of the U.S. peanut crop is used to make peanut butter, which must have a minimum of 90 percent peanuts.

RICE

Farmers began to grow rice on the Grand Prairie of eastern Arkansas, southeast of Little Rock, in 1904. The Grand Prairie is a broad, flat terrace of older alluvium between the valleys of the White and Arkansas Rivers. The brown silt loam soils of the Grand Prairie are underlain by a heavy plastic clay that holds water on the surface. The land is too poorly drained for cotton, which does not like to get its feet wet, but it is excellent for rice because the clay subsoil prevents loss of water by percolation when the fields are flooded.

Rice is yet another specialty crop that is grown in islands in the South, but unlike tobacco and peanuts, it is not grown on land that once grew cotton,

Pine Forests and Paper

Farmers in the six states of the Old South grew 10.8 million acres of cotton in 1949, but in 1982 they grew only 2.3 million acres. What they have done with the 8 million acres of land they no longer use for cotton is to use some of it for other crops, such as soybeans and wheat, or for pastures of highly variable quality that are grazed by cattle of equally variable quality. They have sold or leased some of it to companies that produce forest products., such as pulp and paper. Trees, as far as the cotton farmer was concerned, were merely weeds that took over fallow fields and had to be cleared before the land could be used once again to grow crops, but the wood-using industries have capitalized on the "bonanza forest," forest that grew naturally on the abandoned cotton fields of the South. Wood processing has become a major growth industry in the sometime Cotton Belt. The number of pulp and paper mills in the region has expanded tremendously.

A Personal History

FORREST BLAND

RUNNING A BUSTLING HOUSE OF BROILERS

---◆◇◆---

Forrest Bland at 58 is leathery and lean as a whip. When we first entered his poultry house, I gasped and choked at the acrid smell of ammonia—"That's the smell of money," Forrest told me—but after a few minutes I hardly noticed it.

The floor of the broiler house was alive with clusters of baby chicks that had just been delivered, jostling, pushing, shoving, running, falling, cheeping balls of curious yellow fluff. A metal feeder trough ran the length of the floor. Along either side of the ceiling were rows of bright white lights that were kept on all night to encourage the chickens to eat. The interior was a piano-wire jungle. "Everything is on wires because we have to raise it off the floor with a hand winch before the catchers come to get the chickens. My wife Ruth and I started off with nothing, and we have built it all up. I was a tail gunner on a Marine SBD Dauntless dive bomber in the Pacific during World War II. When that old airplane went into its bombing dive, I sure wished I was back home in Arkansas! I saved all of my money, and Ruth saved all of hers. After the war was over, we bought forty acres and a herd of milk cows. In 1965 or 1966, I just wasn't getting along as good as I wanted in the dairy business, and I started talking to the integrators. Tyson seemed to offer the best deal, and we've stayed with them ever since.

"THAT'S THE SMELL OF MONEY."
[REFERRING TO THE ACRID SMELL OF AMMONIA IN THE CHICKEN HOUSE]

"We started off with one old broiler house. I milked the cows and Ruth tended the chickens. We have to feed and water new chicks the first five or six days by hand from pans we set on the floor, but after that it is all automatic. We walk each house every day to keep the birds stirred up and to check the feed, water, heat, and ventilation to be sure everything is working. The temperature is especially important because the food goes to body heat, not meat, if the birds are too cold, and a hot, humid day really melts 'em down," Forrest said. "We have a water fogger in the roof that puts out spray to cool them in summer.

"These old fields here were all raw and eroded and red before we started putting chicken litter on them to build them up. I overdid it once. I lost 25 cows from the milking herd by feeding them hay that had four times the nitrogen they needed. We made the *Progressive Farmer* on that one! They ran a story as a warning to other farmers. We had used litter plus chemical fertilizers, and the hay was strong enough to kill a cow.

"A few years ago we had a drought, and we couldn't get decent hay. We had to pay $3 a bale for some old worthless stuff, so we began using chicken litter as a supplement. The cattle liked it so much, we've been using it ever since. One-third grain and two-thirds litter is a fine cattle ration." ◆

because it needs a very different kind of environment.

Rice farming is big business. Today's principal rice-producing areas in the South are on the Coastal Prairies along the Gulf Coast of Texas and Louisiana and in the Delta area of eastern Arkansas and adjacent states. The grassy prairies of southwestern Louisiana and Texas were used only for cattle ranches until railroads were built through the area in the late 1880s. The railroads brought farmers from the three "I" states (Iowa, Illinois, and Indiana) into southwestern Louisiana. They rented land from the local cattlemen and started to grow rice just as they had grown wheat in the Midwest, in large fields, with large machines and large grain elevators in the small towns along the railroads. They

grew rice for two years, then put the land in improved pasture for beef cattle for four years, before growing rice on it again.

Farmers from Louisiana carried rice cultivation westward into southeastern Texas, where the flat, low-lying, heavy soils were difficult and expensive to drain. Canal companies acquired large tracts of land, developed extensive drainage systems, sold irrigation water, and rented land to rice farmers. The Texas Coastal Bend area southwest of Houston has the largest rice farms in the South.

STUTTGART

Stuttgart, Arkansas, is in the heart of the Grand Prairie. The rice-drying-and-processing plants in Stuttgart have an

assemblage of grain elevators that can be matched by few places anywhere in the world. The telephone directory lists thity-four aerial spraying services and nineteen chemical companies that serve the needs of rice farmers.

The countryside around Stuttgart is flat and empty. Low earthen levees enclose the fields and snake across them, but there are no fences. The only trees are near farmsteads, which are a mile or more apart because the farms are so large. The farmsteads are clusters of hulking, angular, corrugated metal sheds, some no more than a roof supported by sturdy poles. Each farmstead is a veritable machinery depot, with tractors, trucks, plows, scrapers, combines, and irrigation gear all over the place. Here and there in the fields a

PROSPECTIVE RICE FARMERS evaluate and work reclaimed prairie lands with a gasoline plow and a gasoline road machine, Louisiana, circa 1912. The flat and empty landscape of the prairie provides optimum terrain for rice farming.

(Bottom) A corrugated tin roof provides shelter for a wellhead on The Buffalo River Valley Farm, Arkansas.

(Overleaf) After flowering, mustard produces a pod filled with mustard seed. Mustard, a crop in Georgia, is sometimes used for bee grazing after it has been harvested.

low corrugated-metal roof shelters a wellhead, where a powerful diesel engine pumps water for irrigation into long, straight surface ditches or underground pipes.

The countryside is dotted with reservoirs of various sizes and shapes. Most farmers rotate rice with two years of soybeans, but some rotate rice and reservoirs. After a year or two of rice, they flood the field for a year and stock it with catfish and bass, but fish droppings may enrich the soil so much that the next crop of rice is susceptible to being beaten down by heavy rain and strong winds, and it is very difficult to harvest.

SUGARCANE

Sugarcane is yet another traditional specialty crop of the South. It is largely grown in southern Louisiana because the French-speaking people who began to settle in the area after 1700 clearly understood the facts of life on floodplains. They knew that a river creates its bottomlands when it floods and that the floodplain belongs to the river. People can rarely resist the temptation to cultivate the fertile alluvium that the river has deposited on its floodplain, but they do so at their own risk. Sooner or later the river will rise again and overflow its banks—despite the best efforts of people

to control it with dikes and levees—and sweep away the fruit of their labor.

The highest and best-drained parts of the bottomlands are the natural levees (the banks right next to the stream), where the river dumps the greatest part of its load of sediment each time it overflows. The land slopes ever so gently away from the natural levees toward the backswamps, which even today are still heavily wooded because they are so difficult to drain. The French divided the land into long narrow strips, more or less at right angles to the river, to give each settler some of the better levee land and some of the poorly drained backswamp. The customary farm was a long lot, at

SUGARCANE IS A TALL, thick-stemmed tropical grass with long, drooping leaves. It needs 12 to 15 months to reach full maturity, but farmers in Louisiana must take into account a killing frost between Christmas and Valentine's Day.

(Left) It's an old farmer's saw to say, "It's safe to plant rice when the buds on a pecan tree are as big as a squirrel's ear." Unless there is enough rain to keep a field moist, rice plantations must be flooded to minimize loss of nitrogen and to control weeds.

least eight times as deep as its frontage on the river.

The land is so flat and so low-lying that the settlers had to drain it before they could cultivate it, and farmers still must struggle constantly to maintain their drains. Long, straight, open ditches slice across the fields to carry water from the levees down to the back-swamps. Smaller ditches, called quarter drains, join the main drains at right angles. The quarter drains divide the fields into "squares" or "cuts." The farmer must clean out his ditches each time he cultivates his fields, and he must battle the rank growth of weeds that flourishes along them.

The first white settlers in southern Louisiana, who came directly from France or from the French West Indies, developed large plantations along the Mississippi River as far north as Baton Rouge. They planted sugarcane as early as 1742, but sugarcane did not become an important crop until after 1794. The planters copied the sugar-making system of the West Indies almost exactly, planting the same kind of cane, modeling their mills after those in the West Indies, importing slaves to work the land, and even building the same kinds of houses.

The Cajun settlers put their roads on higher, drier land along the winding

levees, and they built their houses next to the road and the river. They had large families, and their descendants divided their farms down the middle, at right angles to the river, making it narrower and narrower over the generations. Many farms are little wider than the farmstead, but one to three miles deep. The houses are crowded so close together in continuous strips along the levees that southern Louisiana seems to have some of the longest streets in the world.

During the late nineteenth century planters gradually pushed sugar production northward toward the Delta, especially in the years when cotton prices were low. In 1982, sugarcane production was concentrated west of New Orleans, south of Baton Rouge, and east of Lafayette on the natural levees along the Bayou Lafourche, the Bayou Teche, and the south bank of the Mississippi River. In Louisiana they call this part of the state the Sugar Bowl. Farmers grow sugarcane on more than two-thirds of the cultivated land in the Sugar Bowl, and cane is the only crop of any significant consequence. ◆

LABORERS PERFORM the strenuous job of cutting sugarcane, circa 1920. The stalks must be processed as quickly as possible because their sucrose begins to break down immediately upon being cut.

(Right) Many plantation houses were a bit removed from the hurly-burly of daily operations, but still close enough to ensure close supervision.

Rice

The rice plant is a native of tropical marshlands, and it likes a hot, moist environment. It needs high temperatures during the growing season and abundant fresh water for irrigation. The crop needs land that is flat enough to be irrigated but has a slope of two to three feet to the mile so it can be drained fairly quickly. The subsoil must be impermeable enough to avoid excessive water loss by seepage. The rice farmer begins his crop year by leveling off his fields with a giant scraper called a land plane. After he has smoothed a field, he divides it into "cuts" by using a special tractor-drawn plow to "pull" levees 50 to 100 feet apart that snake along the contours.

Two-thirds of the American rice crop is exported. Half the world's people live on rice, but Americans do not like it very much. In 1982 the average American consumed 114 pounds of wheat flour, 65 pounds of potatoes, and only 12 pounds of rice. We consumed a quarter of our rice in the form of beer. We ate about 60 percent directly, and the rest was processed into soups, cereals, and baby food.

The farmer may plant rice seed either on dry ground or in a flooded field. A dry field must be flushed (flooded and drained immediately) after it has been planted unless there is enough rain to keep the soil moist. The levees obviously must be built before an airplane can plant seed in a flooded field. A wet-seeded field must be drained to parch aquatic weeds as soon as the rice plants stick their heads above water. A rice farmer normally floods his fields two to four weeks after seeding, when the plants are six to eight inches high. Flooding a field helps to minimize loss of nitrogen and to control weeds, although a farmer might have to drain it temporarily in order to get into it with ground machinery if it becomes infested with diseases or insects.

Every acre of rice grown in the United States is irrigated. Rice farmers drain their fields and harvest their crops with self-propelled combines when the stems start to droop under the weight of the seedheads. At harvest the grain contains 18 to 22 percent moisture, and the farmer must reduce the moisture content to 12 percent as soon as possible to prevent spoilage. Many farmers have their own grain-drying facilities, but some still haul their rice to a commercial drier to be dried and stored. Enormous batteries of elevators for storing rice dominate the small towns in rice areas. ◆

SCIENTISTS DIVIDE FIELDS OF RICE into experimental patches where different varieties are innovated and tested to enhance quality and quantity.

(Bottom) The ground starts to take on a darkened color as rice fields become flooded from underground pipes.

ENORMOUS BATTERIES OF GLEAMING white elevators used to store rice dominate the landscape of small towns in rice areas of rural America.

(Bottom) When the stems of the rice plants start to droop under the weight of the seedheads, rice farmers drain their fields and harvest their crops with self-propelled combines.

FROM CITRUS TO SUGARCANE

NEW IRRIGATION SYSTEMS have helped to tranform part of the Everglades from a wild, inhospitable swamp into a fertile garden with rich harvests of sugarcane, radishes, lettuce, and sweet corn, and other vegetables.

ost Americans probably think of oranges when they think of Florida, and rightly so, because the state produces three-quarters of the nation's supply, and oranges are its most valuable agricultural product. Oranges were introduced to Florida by Spaniards, who planted citrus trees soon after they founded St. Augustine in 1565. Seminole Indians carried fruit from the Spanish settlements to all parts of the peninsula, and they scattered the seeds so widely that later visitors assumed, quite incorrectly, that oranges were native to the state. ◆ The first commercial orange groves were planted in 1819 in the north-eastern part of the state near navigable waterways such as the St. Johns River. The early groves were small. The fruit was

packed in barrels padded with Spanish moss and shipped in slow sailing schooners to ports in the Northeast. Oranges arrived during the winter holiday season when other fruits were in short supply. They were such an expensive luxury that many children saw them only once a year, when they found oranges tucked in the toes of their Christmas stockings.

In the winter of 1894–95 severe frosts devastated the citrus groves of northeastern Florida. Most of the owners replanted their groves immediately, but in 1899 another extremely severe freeze killed most of the tender young trees. The citrus business in the area never recovered. Growers moved south and began to develop new groves on the warm, sandy ridge down the middle of

the state that had been made accessible by the completion of a railroad from Jacksonville to Tampa in 1884.

The central ridge is the principal citrus-producing area in Florida. The greatest concentration of groves are found west of Orlando and east of Lakeland. U.S. 27, the Citrus Trail, passes through close to two-fifths of the nation's citrus groves between Ocala in the north and Frostproof in the south. Another fifth are near U.S. 1 north and south of Fort Pierce in the Indian River country along the Atlantic coast. Most Florida oranges are processed into frozen juice concentrate, but the Indian River area specializes in premium-quality fresh fruit for direct consumption. It also has one-third of the nation's grapefruit trees. The warm, moist climate of

Florida is best suited to producing juice fruits, such as oranges, grapefruit, and tangerines, but lemons do better in the hot, dry areas of California and Arizona.

The citrus ridge is the backbone of Florida. During the recent geological past, when sea level was several hundred feet higher than it is today, the limestone ridge was covered by thick deposits of sandy material. The porous, gray, sandy topsoil is extremely droughty, and it contains almost no major plant nutrients. "This citrus land is so high and dry that it's not much good for anything else," said Robert E. Norris, a former county agent in Lake County, Florida. Lake County is one of the state's leading citrus counties.

"We produce fruit from fertilizer, not from the natural fertility of the soil.

"THE LAND OF LAKES AND CITRUS
GROVES," Haines City, Fla., circa 1919.
Like some tropical paradise, the
Florida landscape is covered with
miles and miles of rows of orange
and grapefruit orchards.

(Right) An orange pickers' hut in the
Everglades, circa 1899.

149

The soil doesn't do much more than hold up the trees, and we supply all the plant food they need. We select grove sites for frost protection, not for the fertility of the soil, because it doesn't have any to speak of."

Citrus-grove owners start their groves with young trees that have been top-grafted at a nursery. Spring is the time to transplant. A young grove should be fertilized every six weeks, and older groves three times a year. All groves must also be sprayed for insects and diseases at least three times a year. A grove needs a summer cover crop to protect the light, sandy soil from the heat of the sun and from erosion by torrential thunderstorms. In the fall, the grove owner plows under the cover crop to add organic matter to the soil and to allow the bare soil to absorb heat from the sun. The young trees should start to bear fruit by the fourth year, and they should begin to pay for themselves by the eighth.

Picking starts in October. A busy season in December and January is followed by a lull in March and then another peak season in May. Most of the fruit is picked by crews on a piecework basis. In the morning, as soon as the dew has dried, they lean lightweight ladders against the trees, clip the fruit from the branches, and put it in big bags that are slung over their shoulders. When the bag is full, they empty it into a field box. A tractor hauls the full field boxes to the edge of the grove, where the

YOUNG CITRUS TREES must be irrigated and fertilized every six weeks, and all groves must be sprayed for insects and diseases.

(Left) The warm, moist climate of Florida is best suited to producing juice fruits, such as oranges, grapefruits, and tangerines, but lemons do better in hot, dry areas.

(Opposite) Leaning on ladders balanced against the trees, citrus laborers pick oranges from seemingly endless rows of an enormous orchard.

fruit is transferred to a truck or trailer that carries it to the packinghouse.

Before World War II most of the fruit was shipped fresh, but by 1960 more than three-quarters of the oranges and half of the grapefruits were processed into canned juices, segments, or frozen concentrates. Production of frozen orange-juice concentrate expanded.

Florida is citrus country. Today, growers have formed cooperatives that enable a family to retain the ownership of a small grove, even when they cannot afford the equipment necessary to care for it properly. Is it the wave of the future in American agriculture? Though there

are many problems with this system, only time will tell.

For now, the question remains: Can Florida compete with Brazil, which began large-scale production of oranges around 1980? Brazil now produces as much citrus fruit as the United States and exports most of it as concentrate. American consumers used to resign themselves to having to pay more for their morning orange juice every time they read about a freeze in Florida, but Brazil has begun to fill the gap. Recently the price of orange juice has risen for only a week or two after a Florida freeze before concentrate from Brazil starts

flooding in to lower it again. The reality has changed, and Florida is not the domestic or world supply source that it once was for fresh or processed citrus.

THE EVERGLADES

The Everglades area south and east of Lake Okeechobee in southern Florida—among other things—is another major sugarcane-producing district on the mainland of the United States. The Everglades are the southern half of a gentle trough of low-lying, poorly drained swamp and marshland that reaches down the center of the state from

151

WATER MANAGEMENT in the Everglades. This 20-mile pump station is the largest pump station in the world.

(Left) Vast seas of gently waving saw grass, which stands in water for half the year or more, once covered a huge area of the Everglades. Today much of that saw grass has been replaced by richly fertile land.

Lake Tohopekaliga just south of Orlando to the Gulf of Mexico.

Lake Okeechobee is in a large, shallow basin near the center of the trough. In its natural state the lake has no clearly defined shoreline. The water at the edge was too deep for plant growth. The lake simply overflowed along much of its southern rim, and a sheet of water 40 to 70 miles wide seeped southward through the Everglades to the Gulf of Mexico. The land slopes no more than two or three inches to the mile, and the water oozed along at a rate of only about 20 feet a day. It was only a few inches deep during the winter dry season, but heavy rains in summer and early fall could raise it as much as two feet or more.

Southern Florida is one of the wettest areas in the United States. The rainfall varies greatly from season to season, but it averages about 60 inches a year.

Winters are mild and relatively dry. The alternation of wet summers and dry winters has encouraged the growth of saw grass, a tough, coarse plant that is not really a grass at all, but a sedge that grows six to ten feet high. A vast sea of gently waving saw grass, which stands in water for half the year or more, once covered the entire area. The Seminole Indians, who took refuge in the area and learned to navigate it in their canoes, called it the Everglades, "the river of grass."

The remains of the saw-grass plants gradually accumulated when they died

153

and fell into the water because the normal microorganisms of decay require oxygen, and they cannot work on vegetative matter that is underwater. The residues of partially decomposed marsh plants in the Everglades form one of the world's largest deposits of organic—peat and muck—soils.

Organic soils can be extremely productive if they are drained. The soils of the Everglades are especially good because they have been formed in a basin that is floored with limestone, so they have a high lime content. They lack copper, manganese, zinc, and other trace

elements that are essential for plant growth, but the trace elements are needed only in minute quantities that are easy to supply.

In 1907 the state of Florida dug five major drainage canals in the Everglades. Private landowners have crisscrossed the area between them with a herringbone of lesser canals, drains, and ditches. Water percolates laterally through the soil both to and from the ditches, and farmers control the water level in their ditches to irrigate as well as to drain. They pump water out of the ditches and into the canals to drain the land during the rainy

A LONE PALM TREE stands tall above this vista of the Everglades area, circa 1920. The canals snaking through the land are used for drainage.

A Personal History

Ray Roth

Agriculture That Is Dependent on Marketing

Ray Roth had built up a successful vegetable farm from scratch. "I was born in Cleveland in 1924. My father had a truck farm that was annexed into the city before I started school. He grew just about every kind of vegetable you can think of, so I learned how to grow everything. When I got out of the service, I wanted to build a greenhouse, but I couldn't borrow the money. In 1948 I came down here to visit a service buddy, and the next year I came back and sharecropped with his boss. I had ten acres of endive and escarole."

Ray and his partners picked up the option to buy land when the Cubans left to join the Bay of Pigs operation.

"The price of land depends on, among other things, how long it has been cultivated, because this organic soil has a life, and some day it just ain't gonna be there no more."

In summer Ray Roth has to flood the vegetable fields to keep down weeds and to reduce soil loss by oxidation. About ten years ago he started growing rice in the flooded fields as a summer cover crop; he plowed it under in the fall to add fiber to the soil, which it does not get from vegetables.

"Of course I spend time on the farm, but I'm not really a farmer any more. I have an outstanding manager, and he runs the farm. I spend a lot of time trying to keep finance lined up, and I spend time at the packinghouse, talking to the salesmen, finding out what they are able to sell. One of the most important things in agriculture today is marketing, and I attribute our success to the skill of our marketing organization." In the 1950s he joined the South Bay Growers Co-op, which was formed to give growers a more rounded package for marketing.

"The cost of harvesting is the only thing that matters in the produce business. You've already paid the cost of seed and fertilizer before you can harvest the crop, but harvesting is the big expense. You don't harvest a crop unless you can sell it for enough to break even. If the price is too low, you just plow it under and plant the block again. A lot of growers figure they're doing all right if they harvest two-thirds of what they plant, but over the years we have left less than 10 percent of our crops in the field. None of my crops can be frozen or processed. They must be sold immediately.

"Vegetables are very easy to over-produce. This is a major supply area, so a disaster here is reflected in our prices. Of course, a good year here can also hurt you because everybody has an abundant supply, lowering the price. We have absolutely no government interference whatsoever—no quotas, no allotments, no price supports, nothing like that. The vegetable growers in Florida like it that way!" ◆

"None of my crops can be frozen or processed. They must be sold immediately."

season, from June through October, and they can pump water from the canals into their ditches to provide water for irrigation during the winter dry season.

In 1928 a disastrous hurricane struck the area, causing a major disaster to people and crops. Lake Okeechobee is less than 20 feet deep, and steady, galeforce winds from the north blew the northern end of the lake completely dry. The winds drove the water to the southern end of the lake, where they lashed it into waves more than ten feet high. These waves battered and finally breached the low muck levees, and the water poured through in a devastating flood that drowned more than 1,800 people. The U.S. Corps of Engineers was immediately called in to build a sturdy new dike 38 feet high around the lake. The canals drain the agricultural area, and the dike keeps the lake from overflowing and flooding it over again.

The construction of drainage canals had lowered the water level in the Everglades enough to enable farmers to start growing winter vegetables on a large scale, but it has also generated a whole new set of problems. As soon as the soil is exposed to air, it begins to subside, and microorganisms of decay attack the vegetative matter. Also, the soil is so spongy that it is easily compressed by heavy farm machines. After a long dry spell it may even catch fire and smolder for days until it is soaked by a heavy rain. The soil is disappearing at a rate of about an inch a year, and soil subsidence may force the abandonment of much of the agricultural land by the year 2020.

Farmers can minimize subsidence by keeping the water table as high as

The most important crops in the Everglades are sugarcane and vegetables. Some vegetable farmers grow sugarcane as a sideline crop, but more than two-thirds of the cane is produced by four large corporations that grind the cane in their own mills.

The acreage of sugarcane in the Everglades was fairly small until 1960, when the United States placed an embargo on imports of Cuban sugar after Castro took over. Farmers had grown sugarcane in the Everglades on a small scale in the 1920s, but as late as 1959 the Everglades had only 47,000 acres of cane. Five years later the sugarcane area had increased to 214,000 acres, and in 1982 it was up to 344,000 acres, just about half of the national total.

HARVESTING SUGARCANE

A MEASURING STICK PLACED in the muck of the soil of the Everglades when the land was first drained shows that the top three feet of soil have already been eroded.

(Left) Bean plants and the bean poles that support them look like a field of tepees.

possible when the soil is cultivated and by flooding it when it is not. They like to keep the water table low, however, because a water table that is too low harms their crops less than one that is too high, and a low water table is also a precaution against sudden heavy rains. Pasture is the most protective form of land use because a sod cover and a high water table slow the rate of subsidence, but pasture also gives the lowest returns. Most farmers in the Everglades will tell you that their land is too valuable to be "wasted" under pasture.

The sugarcane harvest in the Everglades runs from November through April. The rich, black muck soils support such lush growth that the mature stalks curve in all directions instead of standing erect, and no one has been able to develop a machine that can harvest the tangled stalks satisfactorily. The best machine available is self-propelled on caterpillar treads. Long arms protrude awkwardly to the front and one side. The front arm has revolving blades that slice off the tops of the stalks. The machine then grabs the stalks, cuts them into short

157

lengths, and puts them on a conveyor belt in the side arm that transfers them to a trailing field wagon pulled by a four-wheel-drive tractor.

Cane growers harvest as much as they can by machine, but 70 percent of the crop still has to be cut by hand. A field is fired the day before it is cut to burn off the dry dead leaves and other trash and to drive out any poisonous snakes before the cutters start work.

The workers cut the stalks with vicious-looking machetes, hack off the tops and any remaining leaves, and toss the stalks behind them. After the field has been cut, a self-propelled loader comes rumbling in on caterpillar treads, scoops up the stalks, and loads them into angular field wagons with wire-mesh sides that can hold four tons of cane. Four-wheel-drive tractors haul trains of four field wagons to a transfer ramp near the field. At the ramp, each wagon in turn is tilted to dump its contents onto an angled conveyor belt that loads the stalks onto the massive cane trucks that haul twenty-ton loads of cane to the mill to be ground.

WINTER VEGETABLES

Winter vegetables, including sweet corn; leafy vegetables, such as celery, lettuce, escarole, endive, parsley, cabbage, and cauliflower; and root-type vegetables, such as radishes, carrots, and onions, are second only to sugar as a

THIS FIELD OF YOUNG PEPPER PLANTS is an example of the kind of vegetable farming that has been made possible by land irrigation and drainage in Florida.

(Left, top) The sugarcane harvest begins in the middle of October and continues through December. The dry leaves are burned, filling the air with a heavy caramel smell.

(Left, bottom) Vegetable farms extend throughout the wide-open spaces.

source of farm income in the area. The farmers on the sandy land over near the coast grow fruit-type vegetables, such as peppers, tomatoes, eggplant, beans, and squash. This is made possible by the mildest winters on the mainland of the United States.

Farmers in the Everglades grow more than forty different varieties of vegetable crops. They start planting in September, and are still harvesting in April or May. Most farmers grow only one crop a year, but the vegetable growers produce several crops a year on the same land.

The vegetable land is fallow from May through August because the summers are too hot and too wet for growing vegetables. Vegetables can tolerate heat, and they can tolerate water, but not both at the same time. The combination of 85°F temperatures and soggy feet will steam them to death. The growers flood their fields after the last harvest to reduce oxidation of the soil and to control pests. ◆

Citrus frost

Summers in central Florida are hot and muggy, but the winters are pleasant. The average daytime temperature in January runs around 70°F, but every winter has a couple of cold spells of two or three nights each. A freeze is caused by a cold, dry air mass from the north. The first night, when the cold front passes, is fairly windy, but the second night is cold and still, with clear skies. On such cold, clear nights the soil radiates heat to the atmosphere and cools rapidly. The cold soil chills the air next to it, creating a temperature inversion. The layer of air near the ground is colder and heavier than the air higher up, and it drains downslope to the lower areas, which are frost pockets. Frost damage is least likely on the ridges and the higher slopes and most likely in the low-lying areas where cold air accumulates.

The worst danger of frost on citrus plants is in the last two weeks of December and the first two weeks of January. Fruit is damaged if the temperature drops to 29°F for four hours, and temperatures of 20°F or lower for eight hours can kill the trees.

Water bodies hold heat much better than the soil. Lakes give frost protection by moderating the temperature of the cold air that crosses them and by warming the cold air that accumulates above them. A slope on the southeastern side of a large, deep lake is the best location for a citrus grove.

In 1983, Florida had a big freeze. There was nothing unusual about that because Florida seems to have a big freeze nearly every winter, and the initial reports of damage are usually pretty exaggerated. There were freezes in 1977, 1978, 1980, 1981, and 1982, for example, and after each one those who enjoy making dire predictions had a field day making dire predictions about the gloomy future of the citrus industry in Florida, but before long it was business as usual once again in the groves.

The Christmas freeze of 1983 was quite another story. It absolutely devastated 200,000 acres of citrus groves, an area one-sixth the size of the entire state of Delaware. Two years later the groves north of Orlando still looked like they had been cauterized by a forest fire. The long, straight lines of trees were rows of gnarled, gray ghosts. Their trunks and branches were gaunt and bare, without sign of leaf or life. The dead black wood showed through cracks where frost had split the bark. Lush green weeds flourished beneath the trees where spraying had been discontinued. An occasional plume of dirty smoke marked the funer-al pyre of a pile of dead trees that a bulldozer had uprooted and heaped for burning. Flocks of snow-white egrets picked over the worms and insects in the soil overturned by the bulldozers.

"Our citrus acreage has been declining slowly anyhow," said Norris, "because central Florida is one of the fastest-growing areas in the United States. Some growers have been bulldozing their groves for urban development, but freezes have hit the citrus business pretty hard. In 1962 'the freeze of the century' dropped us from 27 million boxes to 13 million. We were up to 45 million boxes by 1980, but in 1985 we produced less than half a million. In 1982, Lake County had 117,730 acres of citrus, but in 1985 we only had 12,183 acres, and that was just what was alive, not necessarily bearing. It's going to take us at least ten years to get back to 15 million boxes." ◆

ROWS OF HEATHLY CITRUS orchards still define the Florida countryside.

(Previous page, far right) Fallen oranges cover the ground around an orange tree devastated by frost.

AMBER WAVES
OF GRAIN

THE VAST TREELESS PLAINS of western wheat country give a quiet, majestic beauty to the wide-open spaces that blanket the Prairie States, eastern Oregon, and Washington.

Farming in the West is constrained by water and by topography—not enough of one, too much of the other. Few areas west of the north-south boundary between Texas and Oklahoma (the 100th meridian) have enough precipitation for rain-fed agriculture, and most of those that do are topographically unsuited for crop production. The wettest areas in the West are mountainous. Their lower slopes are too steep to be cultivated, and their upper slopes not only are steep, but they are too cool for nearly all crops. They are critically important, however, because the irrigated areas of the West depend on runoff from the mountains for water. ◈ The majestic and apparently inexorable westward march of American agriculture began to sputter and stall when farmers pushed into the semiarid environment of the Great Plains, which sweep from

North Dakota southward to Texas. The land is level, the soil is fertile, and it can produce good crops if it is adequately watered, but rainfall is scanty, and it is completely unreliable. An average year has enough rain to raise a crop of wheat, but few years are average. Many a year has been so dry that the farmers could only watch in helpless despair as their crops withered away and the relentless wind blew their precious topsoil eastward in great billowing dust storms.

Farmers on the dry land margin experimented with other crops, but eventually they realized that wheat was the only crop they could grow with any hope of success. Wheat became the dominant crop of the Great Plains by default, just as it is in the other dry lands of the world. The vast treeless plains of wheat country are the eastern margin of the West, where the wide-open spaces begin. Machines and farms both must be big, because the semiarid land yields dividends so grudgingly in return for human effort. Wheat farms are twice the size of farms in the Corn Belt, but less than half the size of cattle ranches, which predominate in areas that are too dry and too dissected even for wheat. In the humid East, farmers measure their land in acres, but wheat farmers in semiarid areas think in terms of quarters (160 acres), and ranchers in the dry lands talk about sections (640 acres, or one square mile).

Wet and dry years seem to run in cycles in wheat country. A series of wet years happened to coincide with

EARLY WHEAT GROWERS bring in the harvest using an elaborate system from Hunter and Thompson's threshing outfit, circa 1909.

(Right) Harvesting on the Caldwell Farm, Brookings, South Dakota, circa 1898. Stacks of wheat covered the countryside of the West then.

FOR CONVENIENCE, many farmers lay out their fields in long, narrow, parallel strips that they can flip-flop each year. The fallow strips of rich, dark, chocolate-colored soil alternating with amber waves of ripening wheat give many parts of wheat country a picturesque striped appearance.

(Left) Oat harvest, Washington.

World War I, when world wheat prices were sky high, and farmers on the Great Plains planted everything in sight. Then came the dry years of the "Dirty Thirties," when scarcely any rain fell at all, and seeds never sprouted. Strong winds whirled across the Dust Bowl of southwestern Kansas and the Oklahoma-Texas Panhandle, swept up the dry black soil, and blasted it eastward in great dust storms that darkened the sky as far east as Washington, D.C., and laid dirt on the decks of ships three hundred miles out in the Atlantic Ocean.

The wheat country of the Great Plains consists of a spring-wheat area centered on North Dakota and Montana and a winter-wheat area centered on Kansas and Oklahoma. Winters in the spring-wheat area are so severe that farmers must wait until spring to plant their seed, and they harvest the crop in August. In the winter-wheat area, farmers can plant in the fall. The seeds lie dormant in the ground over winter, get a jump start in spring, and the crop is ready for harvest by June.

Many winter-wheat farmers run herds of beef cattle on land that is not

suited to cultivation, and they turn cattle onto their wheat fields in spring. Cattle grazing actually helps to improve wheat yields, because a wheat plant responds by "stooling," or putting out multiple new stalks, after a steer nibbles off its tender young stalk.

The third wheat area in the West is the Palouse country of eastern Washington and parts of Oregon and Idaho, which is one of the few areas west of the Great Plains where crops are grown without irrigation. The Palouse is underlain by deposits of ancient wind-blown dust that are up to 200 feet thick. Many slopes are so steep that conventional farm machinery would tip over, and farmers in the Palouse must use low-slung tractors with treads rather than wheels. Steep slopes usually invite erosion, but the Palouse has just barely enough rain to grow wheat, and the rain is so gentle and drizzly that even the steepest slopes are eroded only slightly.

THE IRRIGATED WEST

Irrigation farmers in the West can grow few crops that cannot also be grown—without the expense of irrigation—in areas much closer to the major markets of the East. All oases produce a variety of crops, many for local consumption, but most irrigation farmers and most oases have had to specialize in producing one principal commodity, or at most only a few, that they can produce so efficiently

HUGE, WHITE GRAIN-STORAGE TANKS cover the western countryside. And there's plenty of it: wheat surplus is a historical phenomenon in America.

(Right) The breathtaking Palouse landscape rises into many slopes that are so steep, conventional farm machinery would simply tip over. Farmers must use low-slung tractors with treads rather than wheels to plant and harvest their wheat.

(Overleaf, left) Apple orchards in winter are in a dormant state, which provides the ideal time for pruning trees. Apples are the principal crop of the irrigated oases of the valleys on the eastern flanks of the Cascade Mountains in Washington.

(Overleaf, right) In the spring, apple-blossoms are cropped to control fruit production.

Sugar-beet pulp makes excellent feed for beef cattle and is grown on small acreages in rotation with other crops.

Sugar Beets

The principal early irrigated areas of the Great Plains were in the valleys of the major rivers. Waters from the Arkansas in Colorado and Kansas made sugar beets the leading cash crop. Beet thinning and harvest required much hand labor. The roots of sugar beets must be hauled to processing plants, where the juice is squeezed out of them and boiled down to make sugar. They are so heavy in relation to their value that they cannot be hauled very far, so sugar-beet growers must be near a processing plant. The pulp from which the juice has been extracted makes fine feed for beef cattle, and so are the leafy tops of the plants. Sugar-beet areas have an abundance of cattle because growers must rotate beets with other crops, such as corn and alfalfa, so many sugar-beet areas have feedlots for beef cattle.

and on such a large scale that it can stand the cost of transportation to any grocery store anywhere in the United States.

The northern oases specialize in crops and fruits that are suitably hardy. The Snake River Plain of southern Idaho is renowned for its baking potatoes, and sugar beets are a specialty of the irrigated river valleys of the northern Great Plains.

THE CENTRAL VALLEY

The southern oases of the West specialize in producing tender crops and fruits. The Central Valley of California, 400 miles long and 50 miles wide, is the world's most awesome oasis. It is easy to irrigate because it is filled with deep alluvial deposits eroded from the Sierra Nevada. The farms of the Central Valley produce more than fifty major agricultural products that are chiefly responsible for making California the nation's leading farm state.

At first farmers from the East were baffled by the mediterranean climate of California, where rain falls in the cool months of the year, but the summers are desertlike. They learned to plant winter wheat in the fall and to harvest it in the

(Left) Center pivot irrigation systems have transformed the landscape of the West, carving out circular fields rather than the rectangular fields of furrow-irrigation systems.

175

spring. Farmers still plant wheat around the dry margins of the Central Valley, in the "ring around the bathtub" that cannot be irrigated, but since the 1880s, irrigation by streams from the Sierra Nevada has replaced dry land farming.

Rice is the leading crop in the northern part of the Central Valley on heavy clay soils that are not well suited to other crops, and cotton is important on large farms in the central and southern parts.

The entire Central Valley is dotted with orchards that produce an astonishing variety of nuts and fruits, including oranges. California's first orange groves were in the Los Angeles lowland. Urban growth has squeezed them out, but Ventura County still produces most of the nation's lemons, and precipitous hillsides in San Diego County sport groves of avocados.

Not even the most extensive and profitable farming systems can afford to pay city prices for land and water, and California is one of the few areas in the United States where the conversion of agricultural land to urban uses may be considered a problem. For example, the Santa Clara Valley south of San Francisco Bay once was famous for its plum orchards, but it has been transformed into Silicon Valley by the growth of the electronics industry. Though there's much growth there now, it's not producing any fruits or vegetables. ◆

(Above) Wine grapes. (Right) Vast areas in the dry southern end of the valley are covered with irrigated vineyards that produce table grapes and raisins, but the best wine grapes are produced in the narrow valleys of the Coast Ranges. The Napa Valley north of San Francisco produces some of the world's finest wines. The valleys of the Coast Ranges also produce such specialty crops as lettuce, strawberries, and artichokes.

(Overleaf) This breathtaking vista of the California coast, near Santa Barbara, looks out on lush rolling hills overlooking rows of manicured vineyards on the valley floor.

Irrigation

Water for irrigation is the very lifeblood for farming in the West. More than three-quarters of all cropland west of the Great Plains is irrigated, and even on the Great Plains the farmers have increasingly resorted to irrigation to intensify their crop production since World War II. On the Great Plains, irrigation is a supplement to the natural rainfall, but in most of the rest of the West it is essential. Most of the older irrigated areas in the West are long and narrow because they are along river valleys or at the foot of mountains where they can capture water from melting snow.

THE GRAND COULEE DAM IN EASTERN WASHINGTON IS THE LARGEST CONCRETE DAM IN THE WORLD, AND IT HAS CREATED A LAKE 150 MILES LONG. HOOVER DAM EAST OF LAS VEGAS HAS CREATED LAKE MEAD, WHICH IS 110 MILES LONG AND UP TO EIGHT MILES WIDE. THE ALL-AMERICAN CANAL CROSSES EIGHTY MILES OF THE HOTTEST DESERT IN THE UNITED STATES TO CARRY WATER FROM THE COLORADO RIVER AT YUMA TO THE IMPERIAL VALLEY OF SOUTHERN CALIFORNIA WITHOUT HAVING TO PASS THROUGH MEXICO.

The first irrigators in the West diverted water from streams into ditches that followed the contours along the sides of the valleys. Downstream the ditches were higher and higher above the stream, and farmers could divert water from the ditches to irrigate the land between the ditches and the stream. The irrigated farmland "under the ditch" stood in sharp contrast to the dry rangeland above it.

Irrigators naturally developed the easiest areas first, so each new gravity-fed irrigation project has had to be more ambitious and far more expensive than the last, with ever larger dams and reservoirs, ever more elaborate distribution networks of canals and laterals, and very long aqueducts and tunnels. ◆

RECENT IRRIGATION technology has been vastly improved. Center-pivot sprinkler irrigation systems have transformed the landscape of the West.

(Previous spread, left) Irrigation water from the Colorado River Dome Valley near Yuma.

(Previous spread, right) Over 100 center-pivot sprinklers controlled by a central computer irrigate wheat, alfafa, potatoes, and melons along the Columbia River near Hermiston, Oregon.

The Cattle Ranches of the West

WHEREVER THERE IS good grass and plenty of water, there you'll find a herd of cattle roaming freely over the open range. In this Montana drive, the cattle are moved from Spring Creek to Hay Creek in the Big Belt Mountains.

R anches occupy by far the largest part of the West, but it is also the least productive part. Beef cattle, sheep, and goats, which require good areas of pastureland, thrive in the West, but the available land cannot be used for anything else. ◆ In 1865 southern Texas—where cattle had been breeding unchecked during the Civil War—was described as "the cattle hive of the continent," but the animals were a long way from any market. Some were shipped to East Coast cities by boat, but the long sea voyage wasted them. Then cattlemen got the idea of rounding up hundreds of thousands of cattle and driving them slowly northward to the cow towns at the end of the railroad lines that were snaking westward across the Great Plains.

Some cattlemen drifted their herds northward instead of selling them at the railhead, because they had heard that there was good grass to the north near stream valleys. Because a mature beef animal needs ten gallons of water a day, control of water gave the rancher control of the extensive upland areas between those stream valleys.

The cattle roamed freely over the unfenced open range, and the herds of different owners often mingled. They were rounded up twice a year, once in the spring, when the calves were branded with the marks of their owners, and again in the fall, when some were sorted out for shipment east to the Corn Belt to be fattened for market. Ranchers

measure the carrying capacity of their land in terms of animals per section (square mile), and even a small family ranch needs around sixteen sections, or about 10,000 acres.

Many ranchers near mountains use the alpine meadows above the timberline for summer pasture. Many of these meadows have been made parts of

THE ROOTS OF AMERICAN CATTLE ranching are to be found deep in Spain and Latin America. From Mexico, the cowboy culture spread north into Texas. Pictured here, early cowboys on horseback round up the herd that has just been watered from the stream.

national forests to protect the watersheds that yield water for irrigation, and the U.S. Forest Service rents them to ranchers for very low fees. Two-thirds of the rangeland in the West is held by agencies of the federal government and is rented to ranchers because it is too fragile to be entrusted to private owners, who would be tempted to overgraze it in dry years.

Wheat farmers have pushed ranchers out of the better areas of the West and into areas that are too dry, too rough, too sandy, or too cool for wheat. Animals on ranches pasture on the unimproved native grasses, and the principal management technique is to keep them from grazing too long in any one place in order not to deteriorate the rangeland. In dry

years, ranchers often yield to the temptation to overgraze, however, and undesirable plants such as sagebrush have invaded grasslands on the Great Plains. Sagebrush dominates the northern montane areas, which are drier, rougher, and less intensively grazed. The driest areas of the West, in California, Nevada, and Arizona, are dominated by the creosote bush, which is not even eaten by goats—who can manage in areas that are too poor even for sheep, because they can browse on trees and shrubs; the cre-

osote bush is hardly grazed at all. Sheep raised for wool are common in areas that are too dry for cattle, because they can get much of the moisture they need from the dew on the grass.

Beef cattle are the principal livestock in ranching areas in the West, but herds of sheep trek from the desert fringe to mountain pastures each summer, and back down again each fall. The Edwards Plateau of south-central Texas, where ranchers raise Angora goats for mohair, is the nation's leading goat-producing area.

CATTLE ARE ROUNDED UP twice a year. In the fall, many are shipped east from ranches to feedlots in the Corn Belt to be fattened for market.

(Right) Montana cattle drive.

The roots of American ranching are deep in Spain. Spaniards brought cattle and horses to Mexico. From Mexico the cowboy culture spread north into the lower Rio Grande Valley of southern Texas, which became the cradle of the American range-livestock industry. The great trail drives ended in 1881, but they established the grand romantic image of the cowboy and left an indelible imprint on American culture. Ranching is a lonesome way of life, with large ranches separated from each other by miles. "It's a great life for men and dogs," they say, "but it's hell on women and dogs."

The cowboys who came from Mexico used ropes called lassos or lariats when they worked their cattle with agile, little cow ponies. They wore broad-brimmed cowboy hats called sombreros; leather leggings called chaps protected their legs against spiny chaparral brush; and they wore cowboy boots with pointed toes that slipped easily into stirrups.

Two major innovations, windmills and barbed wire, were responsible for the development of the modern range-livestock industry. Windmills enabled ranchers to use the upland areas between stream valleys. They drilled wells in the uplands, where the wind seems to blow almost constantly, and used windmills to pump water to surface drinking tanks for their animals.

Barbed wire helped ranchers to manage their rangeland better by enabling them to control the quality of their stock through control of breeding.

Better animals require better feed, especially in winter, when the range is covered with snow, and ranchers spend much of their summers putting up hay for winter feed. Some simply cut the native grasses for wild hay, but if possible, they like to irrigate, fertilize, and seed stream bottomlands for hay meadows. ◆

THE SPRAWLING BELL RANCH, San Miguel Co., N.M., circa 1914, seems to cover the entire country like a great desert.

(Right) No longer do great trail drives blaze across the West; they are a part of history. We are left, however, with the grand, romantic image of the cowboy etched into American culture. Pictured on the right is a cattle drive in eastern Oregon.

(Overleaf) Lazy E/L Ranch homesteaded in 1901. The ranch is situated at the base of the Beartooth Mountains.

BELL RANCH. SAN MIQUEL CO. N.M.

INDEX

◆▣◆

194

PHOTOGRAPHY CREDITS

◆◉◆

To the American Farmer

who has fed and clothed us so well

and who has received so

little recognition for having done so.

Thank you.